Ecosystem Goods and Services
from Plantation Forests

Contents

Figures, Tables and Boxes

Figures

Tables

Boxes

Contributors

Name

Professor Jürgen Bauhus

Position

University of Freiburg
Institute of Silviculture
Tennenbacherstr. 4
79106 Freiburg
Germany
juergen.bauhus@waldbau.uni-freiburg.de

Dr Hannes Böttcher

Forestry Program
International Institute for Applied Systems
Analysis
Schlossplatz 1
A-2361 Laxenburg
Austria
bottcher@iiasa.ac.at

Dr Rudolf S. de Groot

Environmental Systems Analysis Group
Wageningen University
PO Box 47
6700 AA Wageningen
The Netherlands
dolf.degroot@wur.nl

Dr Lisa Hoch

Deutsche Gesellschaft für Technische
Zusammenarbeit (GTZ) GmbH
Advisor
SCN Quadra 1 Bloco C Sala 1501
Ed. Brasília Trade Center
70.711-902 Brasília/DF
Lisa.hoch@gtz.de

Dr Markku Kanninen

Center for International Forestry Research (CIFOR)
Environmental Services and Sustainable Use of Forests Program
Bogor
Indonesia
m.kanninen@cgiar.org

Professor Peter J. Kanowski

The Fenner School of Environment & Society
ANU College of Medicine, Biology & Environment
Forestry Building 48, Linnaeus Way
The Australian National University
Canberra ACT 0200
Australia
peter.kanowski@anu.edu.au

Professor Rodney J. Keenan

Department of Forest and Ecosystem Science
The University of Melbourne
221 Bouverie St
Parkville, VIC 3010
Australia
rkeenan@unimelb.edu.au

Dr Marcus Lindner

The European Forest Institute
Head of Programme Forest Ecology and Management
Torikatu 34
80100 Joensuu
Finland
marcus.lindner@efi.int

Julia Maturana

Economics School
Universidad Católica Santo Toribio
Av. Panamericana Norte 855
Chiclayo, Peru
jmaturana@usat.edu.pe

Dr Benno Pokorny

University of Freiburg
Institute of Silviculture
Tennenbacherstr. 4
79106 Freiburg
Germany
benno.pokorny@waldbau.uni-freiburg.de

Dr Joachim Schmerbeck   Lector – German Academic Exchange Service
TERI University
10, Institutional Area
Vasant Kunj
New Delhi 110070
India
jschmerbeck.daad@teriuniversity.ac.in

Dr Peter J. van der Meer   Teamleader Forest Ecosystems
Alterra – Wageningen University & Research
Centre
P.O.Box 47
6700 AA Wageningen
The Netherlands
peter.vandermeer@wur.nl

Dr Albert I. J. M. Van Dijk   CSIRO Land and Water
GPO Box 1666
Canberra ACT 2601
Australia

Foreword

When Dr. Juergen Bauhus invited me to write a foreword for the book "Ecosystem Goods and Services from Plantation Forests", I knew I was about to enjoy a very interesting and motivating reading. All of its eight chapters fulfilled my expectations. This publication, produced by a renowned group of forest researchers and academics presents a fresh, balanced and well documented vision concerning the roles, potential benefits and challenges of planted forests, looking at the potential contribution of these valuable resources to the continuous and enhanced flow of ecosystem goods and services.

The preliminary results of the FRA 2010 confirm the global trend of expanding planted forest areas at an average of 5 million ha per year. And presently, societies are expecting more from planted forests. Information on how to approach forest plantations from a comprehensive set of market and non-market benefits is inadequate, hence the value of this book in filling up this gap.

While the economic value of sustainably managed natural forests may be higher than forest plantations, the latter is a clear option for degraded lands and low-productivity agriculture and livestock areas. The book explains how properly planned and implemented forest plantations can yield wide-ranging benefits beside timber. In the context of developing new mechanisms to generate financial compensation for forest environmental services, such as the A/R CDM and REDD +, this is indeed very timely.

The self-contained chapters of the book allow for easy reading. The book discusses the global perspectives of forest plantations as well as analyses their contribution to carbon sequestration; their possible impacts on watershed management, and the opportunities and state of knowledge on how plantation forests contribute to biodiversity conservation. In presenting quantification and valuation methods of ecosystem services from plantations as well as some silvicultural options to enhancing biodiversity in planted forests, and by critically reviewing outgrower schemes aimed at involving smallholders in forest plantation activities, the book provides valuable guidance to the forest practitioner, the researcher and academician, and anyone with interest in sustainable development at the rural level.

The book features two inspiring chapters on policies to enhance ecosystem goods and services from plantations and key recommendations for sustainable forest plantations. These are of particular interest to the policy maker and directly relevant to the following specific objectives of the International Tropical Timber Agreement (IITA 2006):

"(j) *Encouraging* [ITTO country] *members to support and develop tropical timber reforestation, as well as rehabilitation and restoration of degraded forest land, with due regard for the interests of local communities dependent on forest resources*; and

(q) *Promoting better understanding of the contribution of non-timber forest products and environmental services to the sustainable management of tropical forests with the aim of enhancing the capacity of* [ITTO country] *members to develop strategies to strengthen such contributions in the context of sustainable forest management, and cooperating with relevant institutions and processes to this end*".

In presenting up-to-date and unbiased information and analysis, the book contributes to the work of ITTO and its members in the achievement of the abovementioned objectives, as much as it can benefit other related agreements, institutions and forest policy initiatives worldwide.

The need to include forest plantations in an overall landscape approach is highlighted in the book. Plantation forests are an ever more significant element shaping our present landscapes. These forests constitute a potential source of income as well as other direct and indirect socio-economic and environmental benefits, thus enabling the fulfillment of human needs while promoting the conservation of forest ecosystems.

The book tackles issues that have only been recently brought up to the forestry debate. It assists in identifying further research needs and provides the reader with a clear view of the opportunities and challenges confronting forest plantations. It opens the window for the support of innovative forest management using the landscape approach, for the long term benefit of man, woman and the environment.

I trust the readers will enjoy the book as much as I did.

Emmanuel Ze Meka
Executive Director
International Tropical Timber Organization – ITTO
Yokohama, April 2010

Preface

The recognition that forests provide important environmental benefits is not new. Already in 400 BC Plato recognized that the loss of forests could lead to soil erosion and the disappearance of water springs. Since then the knowledge about the diverse functions of forests has greatly advanced, particularly following the development of modern forestry, which began with the use of forests for early industries in the 18th century, and the establishment of ecology as a scientific discipline in the 20th century. The term 'environmental services' first came into use in the 1970s after which the term 'ecosystem services' was coined. Forest ecosystem goods such as timber, food, fuelwood, fodder, ornamental and medicinal resources, as well as ecosystem services such as carbon sequestration, soil and water regulation and habitat for pollinating species and wildlife, are all vital to human health and livelihoods.

Traditionally, many of the non-timber goods and services from forest ecosystems have been viewed as free benefits to society. Owing to the fact that many of these 'public' goods and services are difficult to quantify, and that no market exists for many of them, their value has seldom been expressed in market prices. This is one of the reasons why these public goods from forests were given little weight in private, public, and corporate decision-making.

For a long time it was commonly accepted that 'regular' forest management would not only provide sustainable yields of timber products but would also offer all other ecosystem goods and services in sufficient quantity and quality. It is now well established that this may not be the case in many situations. Processes such as the Millennium Ecosystem Assessment have made us aware that the many changes to natural ecosystems resulting from their use and exploitation have led to the degradation of a range of ecosystem services, possibly with serious implications for our well-being.

Tree plantations are artificial forests that differ greatly from natural ecosystems in terms of their structure and function. While the area of natural forests is shrinking, tree plantations are expanding at a rapid rate and dominate the landscapes in some regions of the world. Due to the strong focus on wood production in large-scale industrial plantations, much debate has been initiated about the lack of balance in ecosystem goods and services, and about the social and economic costs versus benefits of large-scale tree plantations. However, with the increasing global demand for timber and woody biomass, it is clear that tree plantations are here to stay because they represent a very efficient production system for a much-needed renewable resource.

As Jack Westoby stated in his book, *The Purpose of Forestry*, from 1987: 'Forestry is not about trees, it is about people. And it is about trees only insofar as trees can serve the needs of people.' The International Union for Conservation of Nature (IUCN) has estimated that the needs of more than 1 billion people in rural areas can only be met through the ecosystem goods and services provided by forests. Therefore, we claim that it is timely to ask how, and to what extent, forest plantations can help provide these different ecosystem goods and services. We were interested to know how forest plantations might substitute or augment ecosystem goods and services from native forest, and how they could be designed and managed to optimize the provision of ecosystem goods and services such as provision of habitat, clean water, and carbon sequestration. In addition, we wanted to explore how ecosystem goods and service from plantations may be valued, how their delivery may be promoted through appropriate policies, and how local people may actually benefit from this. In this book, we aim to present the current knowledge on these issues. The perspective that we adopt focuses on the non-timber goods and services as the production of wood from plantations has been dealt with in great depth in other publications.

This book originated from work undertaken as part of a project funded by the European Union on the role of forest plantations in a crowded world. The main ideas and concepts emanating from this project were presented and discussed at an international conference on Planted Forests and Sustainable Development held in October 2006 in Bilbao, Spain. Part of this conference was a scientific forum on 'Ecosystem Goods and Services from Planted Forests', and most keynote presentations from this conference have been developed into the chapters for this book.

The time between the inception and publication of this book enabled the contributors to develop and present new ideas and perspectives on these themes and to incorporate the most recent literature. However, in the interim, some changes in the contributors to the book occurred. In particular, we wish to mention Ian Calder, a world-class hydrologist of international renown, who succumbed to motor-neuron disease in May 2009. He will be remembered in particular for his critical examination of public misconceptions of the hydrological effects of forests. Ian was committed to completing his chapter, even until shortly before his death. We are very grateful to Rodney Keenan and Albert Van Dijk for taking over the discussion of the interactions between planted forests and hydrological cycles at a late stage of development of this book and for providing an equally fitting contribution.

The primary conclusion of our book is that tree plantations can meet the needs of a broad range of stakeholder goups. There is now a large body of knowledge and experience showing that tree plantations, if appropriately planned, designed and managed, can deliver a wide range of ecosystem goods and services on both landscape and stand scales. Importantly, the benefits and impacts of plantations are highly context specific. Therefore, the range of impacts and benefits associated with plantation forests, which typically differ from those of other landscape components such as native forests or agricultural lands, need to be assessed, agreed upon and managed in a landscape context. Tree plantations that provide for a range of ecosystem services are likely to be more complex in design and management on a stand and landscape scale than conventional large-scale monocultures. The chapters of this book discuss how this may be achieved with regard to ecosystem services such as the maintenance of hydrological cycles, provision of habitat for biodiversity, and sequestration of carbon. One of the major environmental benefits of plantations does not actually relate to the plantations themselves, but to the fact that they help reduce the harvesting pressure on native and semi-natural forests.

This body of knowledge about the impacts of plantations and their contributions to ecosystem goods and services provides the foundation for developing governance regimes that are consistent with the principles of sustainable forest management. In accordance with these widely agreed principles, societies have the right to expect that plantation forests will deliver more benefits than costs. However, the manner in which trade-offs between different ecosystems goods and services are reached, and the extent to which different interest groups share the benefits will ultimately remain a value judgement. It is clear, however, that enhancing the provision of ecosystem goods and services from plantation forests is an important facet of realizing the benefits of this increasingly important form of forestry.

With this book, we hope to convince the reader that tree plantations, despite their predominant simplicity in structure and shape, and despite the high intensity of management needed, can play an important role in solving, and mitigating some of the pressing current global problems such as the increasing demand for resources and energy, poverty, climate change, or the loss of biodiversity. However, we also aim to demonstrate that much still needs to be done to optimize the diverse functions of tree plantations to serve the needs of people. We hope that this book will encourage readers to engage in this process.

Jürgen Bauhus, Peter van der Meer, and Markku Kanninen
March 2010

Acronyms

A/R CDM	afforestation and reforestation project activities under CDM
ABARE	Australian Bureau of Agricultural and Resource Economics
C	carbon
Ca	calcium
CCAR	California Climate Action Registry
CDM	Clean Development Mechanism
CERs	Certified Emission Reductions
CEPLAC	Comissão Executiva de Planejamento da Lavoura Cacaueira
CIFOR	Center for International Forestry Research
CO2FIX	a model for quantifying carbon sequestration in forest ecosystems and wood products
COP	Conference of the Parties
DWAF	Department of Water Affairs and Forestry (South Africa)
ECCP	Economic Cross-Cultural Programme
EGS	ecosystem goods and services
ETS	emissions trading scheme
EU	European Union
FAO	Food and Agriculture Organization (of the United Nations)
FSC	Forest Stewardship Council
GPP	gross primary production
Gt	gigaton
hh	harvested per household
ILO	International Labour Organization
IPCC	Intergovernmental Panel on Climate Change
ITTO	International Tropical Timber Organization
IUCN	International Union for Conservation of Nature
LULUCF	land use, land-use change and forestry
M	million
Mha	million hectares
MLA	multidisciplinary landscape assessment
N	nitrogen
NETFOP	NETworking FOrest Plantations
NGO	non-governmental organization

NPP	net primary production
NTFP	non-timber forest product
NWFP	non-wood forest product
OECD	Organisation for Economic Co-operation and Development
P	phosphorus
PDM	pebble distribution method
PES	payment for ecosystem services
PIFFR	Fiscal Incentives for Forestation and Reforestation
REDD	Reduced Emissions from Deforestation and Degradation
t	metric ton
TEV	total economic value
UNCED	United Nations Conference on Environment and Development
UNFCCC	United Nations Framework Convention on Climate Change
UNFF	United Nations Forum on Forests
WCPA	World Commission on Protected Areas
WWF	World Wide Fund for Nature

1
Plantation forests: global perspectives

Markku Kanninen

Introduction

Forest plantations now cover about 140 million hectares (Mha) globally, representing about 4 per cent of the global forest area (FAO, 2005a). In terms of wood production, plantations are much more important than their share of the forest area indicates, and their importance is expected to increase with time. In 2000, plantations supplied one-third of the total demand for industrial roundwood. According to some estimates, approximately half of the global industrial roundwood supply will be provided by plantations and planted forests by the year 2040.

Globalization of markets for forest products and services brings new opportunities and challenges for plantation-based forestry and forest enterprises. In addition to meeting the growing demands of the pulp and paper industry with fibre from fast-growing plantations, there are opportunities for value-added wood products for expanding international and domestic markets. However, in forest plantations that are managed sustainably and competitively, intensive silvicultural interventions are required in order to fully optimize the production of high-quality products.

In addition to production of wood and fibre, forest plantations provide several other ecosystem services, including carbon sequestration, clean water production, regulation of the hydrological cycle, and improvement in the connectivity of landscape mosaics for biodiversity conservation and the alleviation of desertification. It is expected that the relative importance of such services provided by forest plantations will increase in the future.

Basic concepts and definitions
Evolution of concepts and definitions
Forest-related definitions have been used to classify data and information on forest land use and vegetation. In global policy processes dealing with forests,

commonly agreed definitions are needed to reach common understanding among stakeholders on measures included in global policy agreements (Sasaki and Putz, 2009; Putz and Redford, 2010).

The concepts and definitions around forest plantations have undergone several changes since the first global plantation assessment by the Food and Agriculture Organization (FAO) in 1965. In the process of defining modalities for the implementation of the Kyoto Protocol in 1997, the issue of definitions has been full of contrasting views and controversies between the 'forestry' and 'climate change' communities in the definition of afforestation and reforestation (FAO, 2002). In addition, several environmental and social groups have strongly criticized FAO on the concepts and definitions related to planted forests and forest plantations – not mainly because of this broadening of the concept from forest plantations to include planted forests, but because they do not consider these as forests in the first place (World Rainforest Movement, 2007, 2009). The paper by Varmola et al (2005) gives a good overview of this historical development and evolution of concepts and definitions.

A major conceptual change in the definition was the introduction of the concept of 'planted forests' in the early 2000s (CIFOR, 2001; Carle and Holmgren, 2003). 'Planted forests' is a broader concept than 'forest plantations', thus creating a continuum of concepts and definitions based on forests characteristics. It was created to allow a distinction from the 'traditional', usually exotic and mono-specific forest plantations and forest plantings of native species, usually grown in mixed-species systems, mainly in temperate and boreal zones – earlier defined as 'semi-natural forests' (Varmola et al, 2005; Carle and Holmgren, 2008). Planted forests are defined as those forests predominantly composed of trees established through planting and/or after deliberate seeding of native or introduced species (Carle and Holmgren, 2008).

FAO (2006a) defines plantations as 'forests of introduced species and in some cases native species, established through planting or seeding, with few species, even spacing and/or even-aged stands'. This definition includes not only industrial plantations established for the production of biomass and timber, etc., it also includes small-scale home and farm plantations, agroforestry plantations and plantations established to achieve ecological objectives, such as soil protection and wildlife management. This broad definition of plantations is encapsulated in the typology of planted forests provided by the Center for International Forestry Research CIFOR, 2001) (see Box 1.1). The types of plantations are not only distinguished by their different purpose, but also by their spatial scale, management intensity, structure and ownership. In the typology provided by CIFOR (2001), the 'managed secondary forests' can be regarded as a transitional type between plantations and other forest types.

This evolution of concepts and definitions has had its pros and cons. On the positive side, experts argue that the introduction of 'planted forests' is a move towards a more inclusive concept allowing a better reflection on all the

---

Box 1.1 *Typology of planted forests*

Plantation type and purpose	Characteristics
Industrial plantation: timber, biomass, food	Intensively managed forest stands established to provide material for sale locally or outside the immediate region, by planting or/and seeding in the process of afforestation or reforestation. Individual stands or compartments are usually with even age class and regular spacing and of introduced species and/or of one or two indigenous species. Usually either large scale or contributing to one of a few large-scale industrial enterprises in the landscape.
Home and farm plantations: fuelwood, timber, fodder, orchards, forest gardens and other	Managed forest, established for subsistence or local sale by planting or/and seeding in the process of afforestation or reforestation, with even age class and regular spacing. Usually small scale and selling, if at all, in a dispersed market.
Agroforestry plantation: fuelwood, timber, fodder	Managed stands or assemblages of trees established in an agricultural matrix for subsistence or local sale and for their benefits on agricultural production; usually regular and wide spacing or row planting.
Environmental plantations: windbreaks, soil protection and erosion control, wildlife management, site reclamation or amenity	Managed forest stand, established primarily to provide environmental stabilization or amenity value, by planting or/and seeding in the process of afforestation or reforestation, usually with even age class and regular spacing.
Managed secondary forests with planting	Managed forest, where forest composition and productivity is maintained through additional planting or/and seeding.

Source: adopted from CIFOR, 2001

investments in man-made forests, as well as social and environmental concerns around forest plantations and planted forests (Carle and Holmgren, 2008). On the other hand, this broadening of the definition almost doubles the area of concern. According to FAO (2006a), the global area of planted forests was estimated to be around 270Mha, compared to around 140Mha reported as forest plantations.

Definition of forest plantations

Forest plantations, defined as 'forest or other wooded land of introduced species and in some cases native species, established through planting or seeding', are divided into two sub-groups: productive plantations and protective plantations. In turn, they are defined by FAO (2005a) as:

- productive plantation: forest plantations predominantly intended for the provision of wood, fibre and non-wood products; and
- protective plantation: forest plantations predominantly for the provision of services such as the protection of soil and water, rehabilitation of degraded lands, combating desertification, etc.

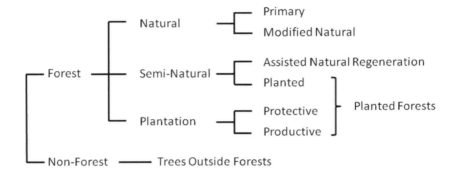

Figure 1.1 *Continuum of forest characteristics and definitions of different forest types according to FAO*

Source: Carle and Holmgren, 2008

Development of plantations

Area of productive and protective plantations

The planted forests thematic study, carried out by FAO, based on the results of the Forest Resources Assessment of 2005 and additional studies, indicated that the total plantation area has increased from 100Mha in 1990 to ca. 140Mha in 2005 (FAO, 2005a, 2006a) (Figure 1.2).

Of total plantation area, 109.3Mha are productive and 30.1Mha are protective plantations. Interestingly, the ratio of productive to protective plantations (3.6) was nearly the same as the ratio between the total area of production to protection forests worldwide (3.7), indicating similar functions. However, for the total forest area, there are other designated functions such as multipurpose or conservation use, not listed for plantations (FAO, 2005a).

China has the largest total area of forest plantations, 31.4Mha in 2005. It is followed by the US, the Russian Federation and Japan, with 17.1, 17.0 and 10.3Mha, respectively. The seven largest plantation countries (China, the US, the Russian Federation, Japan, Sudan, Brazil and Indonesia) account for about

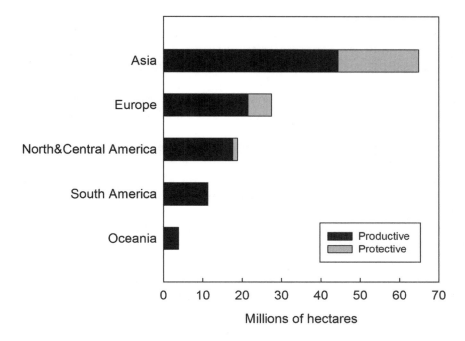

Figure 1.2 *Area of productive forest plantations 1990–2005*

Source: FAO, 2006a

64 per cent of the total plantation area, with a total of 90Mha in 2005. The ten largest plantation countries account for 71 per cent, and the 20 largest countries account for 84 per cent of the world's total plantation area. Productive plantations represented 79 per cent and protective plantations 21 per cent of the total area of plantations, respectively.

China has the largest area of productive forest plantations, about 29Mha in 2005, which is about 26 per cent of the world total. It is followed by the US, the Russian Federation and Brazil, with 17, 12 and 5Mha, respectively. In terms of protective forest plantations, Japan leads with 10Mha in 2005 (35 per cent of the world total), followed by the Russian Federation, China and India with 5, 2.8 and 2.2Mha, respectively. In terms of growth, the fastest growth in plantation area has recently taken place in Asia, where the plantation area has grown by 2.5 per cent per year in recent years.

Countries in which the relative importance of forest plantations corresponds to or approaches 100 per cent of their total forest area (Egypt, Libya, Cape Verde, Bahrain, Kuwait, Oman, United Arab Emirates and Malta) are typically those that have very little natural forest or where the forest area had been diminished in historical times. The latter also applies to countries such as Lesotho, Ruanda, Ireland, the UK, Denmark, Iceland, Israel, Syria and

Burundi, where forest plantation area covers more than 50 per cent of total forest area. The relative importance of plantations in these countries is indicative also of their importance for the selection of ecosystem goods and services, and the selection of these countries shows that plantations play this role in many different parts of the world such as Europe, North and East Africa and the Middle East. Although there is no tight relationship between the percentage forest cover and the percentage that plantations contribute to forest area (Figure 1.3a), it is obvious that countries with the highest plantation percentage have low forest cover. In addition, the highest proportions of protective plantations can be found in countries with low percentages of forest cover. Also, this can be seen as an indication of the substitutive role of plantations, where the cover of native forests has been lost or is naturally low. The absence of a tight relationship also indicates that the relative importance of protective plantations depends on many other factors, such as topography, land-use history and current land use, etc.

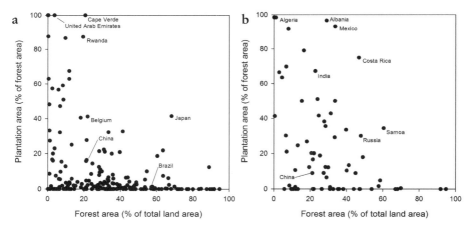

Figure 1.3 *(a) The relative importance of plantations by area in relation to the total forest area, and (b) the relative importance of protective plantations by area in relation to the percentage of total forest areas of countries that have reported protective plantations*

Source: FAO, 2005a

It is interesting to observe that large plantation countries, i.e. countries with large areas of plantations, like the Russian Federation, Brazil, Indonesia and the US, have large areas of natural forests as well, so that plantations represent only a small proportion (1–6 per cent) of their total forest area. In China, the country with the largest total plantation area, the share of plantations is 16 per cent of the total forest area in the country.

About 50 per cent of the total area of planted forests reported in the FAO thematic study (2006a) was under public ownership, whereas plantations owned by smallholders covered 32 per cent and corporations 18 per cent of the total plantation area, respectively. When compared to the situation in 1990, the area (both absolute and relative) of smallholder plantations has tripled, whereas the share of publicly owned plantations has decreased. This is mainly due to the increase in smallholder woodlots and village-scale plantations for the production of wood products for domestic and local use, including firewood and charcoal – a development that started in many developing countries in the 1980s (Evans, 2009).

In terms of the end-use of the products from the plantations, the production of sawlogs dominated with 67Mha (46 per cent of total area) in 2005. The area of plantations aimed for pulpwood and fibre have increased rapidly, covering 18 per cent (27Mha) of the total plantation area reported (Figure 1.4).

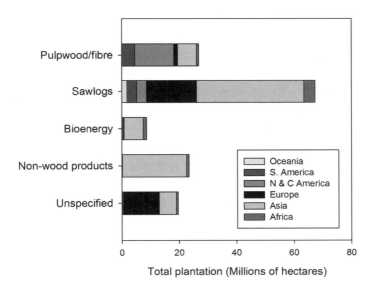

Figure 1.4 *Plantation area (Mha) by the intended end-use 1990–2005*

Source: FAO, 2006a

Species planted

According to the planted forests thematic study of FAO, *Pinus* is the most commonly used genus in plantations, with a total area of 54Mha, representing 29 per cent of the total area planted. The ten most commonly planted genera in plantations, (*Pinus, Cunninghamia, Eucalyptus, Populus, Acacia, Larix, Picea, Tectona, Castanea, Quercus*) represent about 70 per cent of the total area planted (Figure 1.5). In productive plantations, the share of the top ten genera of the total area planted is 77 per cent, whereas in protective plantations the top ten genera represent 60 per cent of the total area planted.

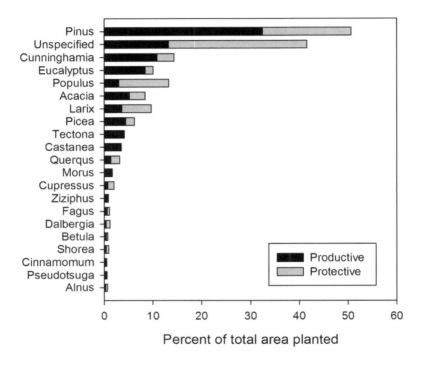

Figure 1.5 *Most commonly used species in plantations*

Source: FAO, 2006a

The pool of species differs to some extent between these two purposes of plantations. The fast-growing plantation species belonging to the genera *Eucalyptus* and *Acacia*, as well as the valuable tropical hardwood teak (*Tectona*) or the *Castanea* grown for non-wood forest products are much less important for protective plantations. Some species that produce litter promoting fires or suppress development of understorey such as eucalypts or teak, are in many situations not particularly suitable for protective functions. In contrast, genera such as *Cryptomeria*, *Chamaecyparis*, *Populus* and *Larix* are relatively more important in protective plantations. However, this difference cannot be attributed to the fact that these species are more suitable for protection purposes than other species. In fact, it may be contested that they are not. The large share of these particular species can be attributed to the country-specific classification of species for productive and protective purposes, which also depends very much on the respective legislation. The importance of the above-listed species for protective plantations is to a large extent attributable to their share of the plantation estate in Japan and China. Particularly in the latter country, large areas of protective plantations were established, or they were established as productive plantations and later declared protection forest (Wenhua, 2004).

The larger proportion of unspecified species in protective plantations may indicate that the species is of lesser importance and/or that the areas under protective plantation may be smaller and more diverse and therefore more difficult to report on. The genus *Pinus* is of great importance for both protection and production plantations.

While fast-growing exotic monocultures dominate plantations aimed at wood or fibre production in the tropics (see Table 1.1), the importance of native species plantations either in pure or mixed stands is gaining importance, particularily for the rehabilitation and restoration of degraded lands (Lamb, 1998; Carnevale and Montagnini, 2002; Gunter et al, 2009). If managed properly, mixed stands offer higher productivity and ecological gains in terms of the provision of multiple ecosystems services compared to pure stands (see Chapter 5; Erskine et al, 2006; Petit and Montagnini, 2006; Piotto, 2008).

The importance of species planted for the provision of biodiversity and hydrological services are discussed in Chapters 4 and 5 of this book. For carbon sequestration services (Chapter 3), short-rotation timber species have high internal rates of return and high carbon benefits. However, the risks to local communities can be lower if long-rotation mixed plantations are used (Hooda et al, 2007). It is interesting to note that of the current 14 A/R CDM projects (afforestation and reforestation project activities under the Clean Development Mechanism) registered (UNFCCC, 2010), about half of the projects use fast-growing trees species, mainly exotics, to achieve maximum carbon sequestration rates, whereas the other half includes several native tree species in their projects; this is to diversify the ecosystem services produced and to maximize benefits to local communities by increasing access to non-timber forest products such as fruits, resins and honey (Ellis, 2003).

Fast-growing plantations

Fast-growing plantations are a sub-group of productive plantations aimed at industrial use with special characteristics in terms of their management. Typically, they are short-rotation plantations with single-species blocks of a high-productive species, usually exotics. They are usually owned by a single company, and they dominate the landscape, constituting the dominant (or only) land use in their area. The dominant objective pursued with these plantations is to produce large volumes of fibre for the pulp industry.

Cossalter and Pye-Smith (2003) estimated that in the early 2000s the total area of fast-growing plantations was about 10Mha (Table 1.1), growing at the pace of about 1Mha per year.

Although the land area covered by fast-growing plantations is relatively small – less than 10 per cent of the total plantation area – these plantations have created a lot of controversy in terms of the social and environmental issues related to them (Cossalter and Pye-Smith, 2003; World Rainforest Movement, 2007, 2009; Evans, 2009).

Table 1.1 *Fast-growing plantations*[1]: *main species and countries involved*

Species	Mean annual increment (m³/ha/year)	Rotation length (years)	Estimated extent as fast-growing plantation only (1000ha)	Main countries (in decreasing order of importance)
Eucalyptus grandis and various eucalypt hybrids	15–40	5–15	± 3700	Brazil, South Africa, Uruguay, India, Congo, Zimbabwe
Other tropical eucalypts	10–20	5–10	± 1550	China, India, Thailand, Vietnam, Madagascar, Myanmar
Temperate eucalypts	5–18	10–15	± 1900	Chile, Portugal, NW Spain, Argentina, Uruguay, South Africa, Australia
Tropical acacias	15–30	7–10	± 1400	Indonesia, China, Malaysia, Vietnam, India, Philippines, Thailand
Caribbean pines	8–20	10–18	± 300	Venezuela
Pinus patula and *P. Elliottii*	15–25	15–18	± 100	Swaziland
Gmelina arborea	12–35	12–20	± 100	Costa Rica, Malaysia, Solomon Islands
Paraserianthes falcataria	15–35	12–20	± 200	Indonesia, Malaysia, Philippines
Poplars	11–30	7–15	± 900	China, India, USA, C and W Europe, Turkey

Fast-growing plantation forests are broadly defined as having average growth rates ranging from ≤10 to ≥40 m³/ha/year, with shorter rotations from ≤6 years to around 35 or 40 years.

Source: Cossalter and Pye-Smith, 2003

Other types of plantations

There are other types of 'tree plantations' that are very common in the everyday life of rural communities in many developing countries, and they are important in terms of their economic value and the ecosystem goods and services they provide. However, they may not be considered to be 'forest plantations' or are not otherwise adequately captured by FAO or by national statistics. These include home gardens (e.g. Méndez et al, 2001; Kumar and Nair, 2006) and other types of agroforestry and silvopastoral systems (e.g. Beer et al, 2000), individual trees grown in rows as wind breaks (e.g. Harvey et al, 2000; Wilkinson and Elevitch, 2000).

The study carried out by FAO on 'trees outside forests' (FAO, 2001, 2005b) gives a good overview of the importance of these plantations. It indicates that it is almost impossible to assess the economic, social or environmental value of these plantations because national statistics do not exist (FAO, 2001). However, since these tree plantings are directly shaped and managed by local people and they grow in their immediate environment, they are particularly important for the provision of ecosystem goods and services.

Plantations and ecosystem services

Increasing role of plantations in wood supply

The share of wood from forest plantations in the total production of industrial roundwood was 5 per cent in 1960, 22 per cent in 1995 and 30 per cent in 2005 (Varmola et al, 2005; Seppälä, 2007). Although the plantations represent only a very small share of the current global forest area, the intensity of management is high. If all industrial wood came from effectively managed planted forests, only some 73Mha, i.e. less than 2 per cent of the world's forest area, would be enough to satisfy the current global need of industrial wood (Seppälä, 2007).

Global industrial wood supply from plantations is increasing rapidly as substantial areas of tree plantations mature. Over the past two decades, most new plantation development has occurred in tropical and subtropical regions and in temperate zones of the southern hemisphere. Industrial timber production from natural forests has begun to decline in the leading tropical forest wood-producing countries of Asia, as supplies of commercially accessible large-diameter timber have fallen sharply in recent years. A report by Spek (2006) on investments in the pulp industry indicates that whereas South America (mainly Brazil and Chile) and Indonesia accounted for more 70 per cent of new investments in the pulp industry in the 1990s, Asian countries (mainly China) dominate the new investments.

Future trends

Global demand for forest products has grown at a rapid pace over the past decade and this is expected to continue in the foreseeable future. In the Asia-Pacific region alone, annual consumption of hardwood pulp is expected to increase by 73 million cubic metres (Mm^3) and annual consumption of softwood pulp by $32Mm^3$. There is a shift in the consumption and production of forest products. Recent studies show that current demand for forest industry products will grow less than before in Organisation for Economic Co-operation and Development (OECD) countries, while at the same time the demand will continue to increase considerably in many developing countries and in countries in transition. This means a shift in the consumption of forest products from Western Europe, North America and Japan to the rest of Asia, Eastern Europe and Russia (Barr and Cossalter, 2004; Seppälä, 2007)

A recent study by Carle and Holmgren (2008) estimated that the area of planted forests will increase from its current level (261Mha) to about 303–345Mha in 2030, depending on the scenario. This increase is 16–32 per cent for the whole 25-year period, or 0.6–1.2 per cent per year. This would mean that the total volume of wood produced annually by planted forests would increase from 1.4 billion m^3 per year in 2005 to a level between 1.6–2.1 billion m^3 per year in 2030.

The above figures are for planted forests and not for forest plantations, which makes it difficult to compare them with earlier results. A study by ABARE-Jaakko Pöyry (1999), reported by Varmola et al (2005), estimated that the industrial wood supply from forest plantations would be 970Mm³ per year in 2020 (44 per cent of the total) and 1.1 billion m^3 per year in 2040 (46 per cent of the total).

In recent years, the voluntary carbon markets have accounted for the bulk of forest carbon transactions: 95 per cent in 2008 and 72 per cent in first two quarters of 2009. In 2008, the total amount of credits traded in the voluntary markets was 5Mt CO_2 and the value US$37 million (Hamilton et al, 2009). Although the share of afforestation and reforestation of the total amount of Certified Emission Reductions (CERs) generated in the Clean Development Mechanism (CDM) is negligible, recent approval of new project methodologies has led to the registration of ten new A/R CDM projects in 2009.

The projections described above both on the supply of wood and carbon credits show that plantations will become a more important part of many landscapes. This development is not uniform in all parts of the world, and an increasingly higher percentage of new plantations will be established in East and South-East Asia and to a lesser extent in South America.

Challenges for the future

The success of forest plantations as sources of industrial roundwood supply has been obvious. In the past 25 years, their share of total industrial roundwood production has increased from 5 per cent to over 30 per cent. In the next 25 years, it is expected to reach 50 per cent of total industrial roundwood production.

This success is often contrasted with social and environmental problems that large-scale plantation schemes have repeatedly caused. In many cases, most smallholders and forest inhabitants have had limited participation – or no engagement at all – in the profitable business developed by large or medium size companies, and they have not shared significant benefits (direct income or other) in the planting, tending and logging operations performed by these companies. In other cases, companies have planted in large uniform blocks of land and this has led to the displacement of local people from the rural villages or small holdings where they have traditionally lived. To improve the situation, certification schemes and other safeguards have been introduced. The voluntary guidelines for the responsible management of planted forests established by FAO (2006b) and ITTO (1993), and by various certification schemes are moves in the right direction. However, there is much still left undone to ensure their general application in practice.

One of the negative impacts of large-scale plantations has been on biodiversity (see Chapter 5). However, this has depended very much on the land use that was replaced by the plantations. If properly planned and managed, plantations can play a role in building connectivity in fragmented landscapes (Kanowski et al, 2005; Marjokorpi, 2006; Marjokorpi and Salo, 2007), or acting as catalysts for native species (Kuusipalo et al, 1995; Keenan et al, 1997; Parrotta and Turnbul, 1997; Otsamo, 2000). But as biodiversity continues to be lost, fixed, single-objective plantation management is likely to be a less attractive option in the future (Lamb, 1998). The plantation management paradigm has to move towards the management of multi-functional landscapes by incorporating multiple ecosystem services as an integral part of the overall production function of forest plantation schemes.

Ecosystem services markets are growing fast and forest plantation will have an important role to play in global carbon markets and, more locally, in payments for ecosystem services schemes for managing water resources. Both of these areas are expected to gain increasing importance in the future, when the actions to combat climate change – both through adaptation and mitigation – become part of the new paradigm of sustainable plantation management.

References

Barr, C. and Cossalter, C. (2004) 'China's development of a plantation-based wood pulp industry: government policies, financial incentives, and investment trends', *International Forestry Review*, vol 6, pp3–4

Beer, J., Ibrahim, M. and Schlönvoigt, A. (2000) 'Timber production in tropical agroforestry systems of Central America', in B. Krishnapillay, E. Soepadmo, N. L. Arshad, A. Wong, S. Appanah, S. Wan Chi, N. Manokaran, H. Lay and K. Kean (eds) *Forest and Society: The Role of Research*, vol 1, pp777–786. 21st IUFRO World Congress, Kuala Lumpur, 7–12 August 2000. Kuala Lumpur, IUFRO World Congress Organizing Committee

Carle, J. and Holmgren, P. (2003) 'Definitions related to planted forests', UNFF Intercessional Expert Meeting on the Role of Planted Forests in Sustainable Forest Management, Wellington, New Zealand, pp329–343

Carle, J. and Holmgren, P. (2008) 'Wood from planted forests: a global outlook 2005–2030', *Forest Products Journal*, vol 58, pp6–18

Carnevale, N. J. and Montagnini, F. (2002) 'Facilitating regeneration of secondary forests with the use of mixed and pure plantations of indigenous tree species', *Forest Ecology and Management*, vol 163, pp217–227

CIFOR (2001) 'Typology of planted forests', *CIFOR InfoBrief*, Center for International Forestry Research (CIFOR), Bogor, Indonesia

Cossalter, C. and Pye-Smith, C. (2003) 'Fast-wood forestry: myths and realities', *Forest Perspectives*, Center for International Forestry Research, Bogor, Indonesia

Ellis, J. (2003) 'Forestry projects lessons learned and implications for CDM modalities', OECD and IEA Information Paper, Paris

Erskine, P. D., Lamb, D. and Bristow, M. (2006) 'Tree species diversity and ecosystem function: can tropical multi-species plantations generate greater productivity?', *Forest Ecology and Management*, vol 233, pp205–210

FAO (2001) 'Trees outside forests: towards rural and urban integrated resources assessment. elements for consideration', Contribution to Forest Resource Assessment 2000 report, Working Paper, Rome

FAO (2002) *Proceedings: Expert Meeting on Harmonizing Forest-related Definitions for Use by Various Stakeholders*, Food and Agriculture Organization, Rome

FAO (2005a) *Global Forest Resource Assessment 2005: Progress Toward Sustainable Forest Management*, FAO Forestry Paper 147, Food and Agriculture Organization, Rome

FAO (2005b) *Trees Outside Forests: Annotated Bibliography*, Food and Agriculture Organization, Rome

FAO (2006a) 'Global planted forests thematic study: results and analysis', by A. Del Lugo, J. Ball and J. Carle, Planted Forests and Trees Working Paper 38, Rome (also available at www.fao.org/forestry/site/10368/en)

FAO (2006b) 'Responsible management of planted forests: voluntary guidelines', Planted Forests and Trees Working Paper 37/E, Rome

Gunter, S., Gonzalez, P., Alvarez, G., Aguirre, N., Palomeque, X., Haubrich, F. and Weber, M. (2009) 'Determinants for successful reforestation of abandoned pastures in the Andes: soil conditions and vegetation cover', *Forest Ecology and Management*, vol 258, pp81–91

Hamilton, K., Sjardin, M., Shapiro, A. and Marcello, T. (2009) *Fortifying the Foundation: State of the Voluntary Carbon Markets 2009*, New Carbon Finance, New York

Harvey, C., Guindon, C. F., Haber, W. A., Hamilton DeRossier, D. and Murray, K. G. (2000) 'The importance of forest patches, isolated trees and agricultural windbreaks for local and regional biodiversity: the case of Costa Rica', in B. Krishnapillay, E. Soepadmo, N. L. Arshad, A. Wong, S. Appanah, S. Wan Chi, N. Manokaran, H. Lay and K. Kean (eds) *Forest and Society: The Role of Research*, vol 1, pp787–798, 21st IUFRO World Congress, Kuala Lumpur, 7–12 August 2000, Kuala Lumpur, IUFRO World Congress Organizing Committee

Hooda, N., Gera, M., Andrasko, K., Sathaye, J., Gupta, M., Vasistha, H., Chandran, M. and Rassaily, S. (2007) 'Community and farm forestry climate mitigation projects: case studies from Uttaranchal, India' *Mitigation and Adaptation Strategies for Global Change*, vol 12, pp1099–1130

ITTO (1993) 'ITTO guidelines for the establishment and sustainable management of planted tropical forests', ITTO Policy Development Series, issue 4

Kanowski, J., Catterall, C. P. and Wardell-Johnson, G. W. (2005) 'Consequences of broadscale timber plantations for biodiversity in cleared rainforest landscapes of tropical and subtropical Australia', *Forest Ecology and Management*, vol 208, pp359–372

Keenan, R., Lamb, D., Woldring, O., Irvine, T. and Jensen, R. (1997) 'Restoration of plant biodiversity beneath tropical tree plantations in Northern Australia', *Forest Ecology and Management*, vol 99, pp117–131

Kumar, B. M. and Nair, P. K. R. (eds) (2006) *Tropical Home Gardens. A Time-Test Example of Sustainable Agroforestry*, Springer Verlag

Kuusipalo, J., Adjers, G., Jafarsidik, Y., Otsamo, A., Tuomela, K. and Vuokko, R. (1995) 'Restoration of natural vegetation in degraded *Imperata cylindra* grasslands: understorey development in forest plantations', *Journal of Vegetation Science*, vol 6, pp205–210

Lamb, D. (1998) 'Large-scale ecological restoration of degraded tropical forest lands: the potential role of timber plantations', *Restoration Ecology*, vol 6, pp271–279

Marjokorpi, A. (2006) 'Biodiversity management in fast-growing plantations: a case study from West Kalimantan, Indonesia', *Annales Universitatis Turkuensis*, Serie AII, Part 200

Marjokorpi, A. and Salo, J. (2007) 'Operational standards and guidelines for biodiversity management in tropical and subtropical forest plantations: how widely do they cover an ecological framework?', *Silva Fennica*, vol 41, pp281–297

Méndez, V. E., Lok, R. and Somarriba, E. (2001) 'Interdisciplinary analysis of homegardens in Nicaragua: micro-zonation, plant use and socioeconomic importance', *Agroforestry Systems*, vol 51, pp85–96

Otsamo, R. (2000) 'Secondary forest regeneration under fast-growing forest plantations on degraded *Imperata cylindra* grasslands', *New Forests*, vol 19, pp69–93

Parrotta, J. and Turnbull, W. (eds) (1997) 'Catalysing native forest regeneration in degraded tropical lands', *Forest Ecology and Management* (Special Issue), vol 99, pp1–290

Petit, B. and Montagnini, F. (2006) 'Growth in pure and mixed plantations of tree species used in reforesting rural areas of the humid region of Costa Rica, Central America', *Forest Ecology and Management*, vol 233, pp338–343

Piotto, D. (2008) 'A meta-analysis comparing tree growth in monocultures and mixed plantations', *Forest Ecology and Management*, vol 255, pp781–786

Putz, F. E. and Redford, K. H. (2010) 'The importance of defining "forest": tropical forest degradation, deforestation, long-term phase shifts, and further transitions', *Biotropica*, vol 42, pp10–20

Sasaki, N. and Putz, F. E. (2009) 'Critical need for new definitions of "forest" and "forest degradation" in global climate change agreements', *Conservation Letters*, vol 2, pp226–232

Seppälä, R. (2007) 'Global forest sector: trends, threats and opportunities', in Freer-Smith, P. H., Broadmeadow, M. S. J. and Lynch, J. M. (eds) *Forestry and Climate Change*, CAB International, pp25–30

Spek, M. (2006) *Financing Pulp Mills: An Appraisal of Risk Assessment and Safeguard Procedures*, Center for International Forestry Research (CIFOR), Bogor, Indonesia

UNFCCC (2010) 'Registered A/R CDM projects' http://cdm.unfccc.int/Projects/registered.html, accessed 12 February 2010

Varmola, M., Gautier, D., Lee, D. K., Montagnini, F. and Saramäki, J. (2005) 'Diversifying functions of planted forests', in Mery, G., Alfaro, R., Kanninen, M. and Lobovikov, M. (eds) *Forests in the Global Balance – Changing Paradigms*, IUFRO World Series Volume 17, International Union of Forest Research Organizations (IUFRO), Helsinki, pp117–136

Wenhua, L. (2004) 'Degradation and restoration of forest ecosystems in China', *Forest Ecology and Management*, vol 201, pp33–41

Wilkinson, M. K. and Elevitch, C. R. (2000) *Multipurpose Windbreaks: Design and Species for Pacific Islands*, Permanent Agriculture Resources, Holualoa, Hawaii, USA

World Rainforest Movement (2007) *WRM Bulletin*, no 117 www.wrm.org.uy/bulletin/117/Bulletin117.pdf

World Rainforest Movement (2009) *WRM Bulletin*, no 146 www.wrm.org.uy/bulletin/146/Bulletin146.pdf

2

Quantifying and valuing goods and services provided by plantation forests

R. S. de Groot and P. J. van der Meer

Introduction

Forests, both natural and planted, can provide multiple benefits to human society, which can be direct or indirect. Direct benefits include goods such as timber, food, fuelwood, fodder, ornamental and medicinal resources, and opportunities for recreation. Indirect benefits comprise services such as carbon sequestration, soil and water regulation and habitat for pollinating species and wildlife (Campos et al, 2005; Chapters 3, 4 and 5). Goods and services provided by forests are a means of livelihood for millions of people; according to the IUCN some 1.6 billion people rely heavily on forests for their livelihoods (IUCN, 2007). However, the value of these goods and services is seldom incorporated in market prices, one of the reasons why they are given too little weight in policy decisions (Costanza et al, 1997). Furthermore, the ongoing, rapid growth of human populations and their resource demands has increased the pressure on forests (and other ecosystems) dramatically over the last decades. In many cases this has resulted in overexploitation and degradation of native forests and a decline in the quantity and quality of the goods and services delivered (Millennium Ecosystem Asssessment (MEA), 2005).

To reduce the pressure on native forests for wood and other resources, plantation forests may be very important yet often they also pose a threat (e.g. when native forests are replaced by oil palm or fast-growing timber plantations). To optimize planning and decision-making about the location and management of plantation forests, it is important to know what goods and services are being provided by the various forests types and forest management systems. Some goods and services are relatively easy to identify and measure, such as the production of timber and some non-timber forest products (NTFPs). Other ecosystem services are more difficult to identify and quantify, like the role forests play in biodiversity conservation, water regulation and cultural services. Once it is known what goods and services are, or can be, provided by which type of forest or management regime, their importance to local communities, and human society in general, needs to be quantified. This can be done in socio-cultural and/or economic terms (including monetary values and employment). The economic valuation of ecosystem goods and services (EGS) is also needed to understand the motives behind different ecosystem management schemes and to evaluate the consequences of alternative courses of action.

In this chapter we review how the importance (value) of forest goods and services can be measured, and how this information can be used to optimize and finance sustainable management of forested areas. We begin with a general typology of EGS for both natural and plantation forests. Then, the main steps and methods for quantifying and valuating EGS are described drawing on a case study in India. In conclusion, we recommend means for the more efficient and sustainable provision of EGS in plantation management.

Ecosystem goods and services of plantation forests

Ecosystem goods and services: the general concept

To understand the capacity of ecosystems (including forests) to provide goods and services of value to human society, an integrated approach is essential. A general framework for such an approach is given in Figure 2.1.

The core of this approach is the translation of ecosystem structures and processes into goods and services with, as a bridging concept, the notion of 'ecosystem functions' which are defined as 'the capacity of natural processes and components to provide goods and services that satisfy human needs, directly or indirectly' (de Groot, 1992). It is important to note that 'ecosystem functions' always exist, whereas 'ecosystem goods and services' are only recognized when people use or benefit from them. For instance, the regulation of surface water flows, a function of all forests, only becomes a service when people are affected by it (e.g. Campos et al, 2005).

The definitions of ecosystem goods and services in the literature vary. Daily (1997) defines ecosystem services as 'the conditions and processes through which natural ecosystems, and the species that make them up, sustain and fulfil human life'. On the other hand, Shelton et al (2001) state that 'ecosystem services have been conceptualized as transformations of natural assets (soil, water, atmosphere and living organisms) into goods and other products (including experiences) that are valuable to people'.

Costanza et al state that 'Ecosystem goods (such as food) and services (such as waste assimilation) represent the benefits human population derives, directly or indirectly, from ecosystem functions' (1997, p253) and apply the term ecosystem services to all benefits arising from ecosystems including goods.

The Millennium Ecosystem Assessment follows a similar and very practical

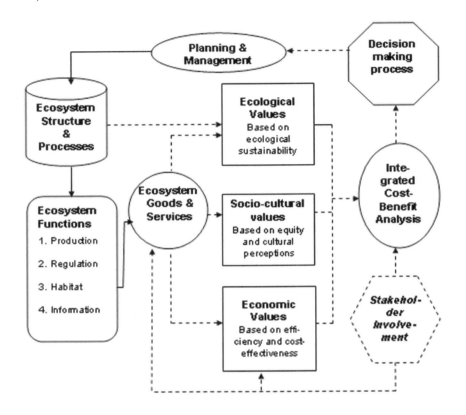

Figure 2.1 *Framework for integrated assessment and valuation of ecosystem functions, goods and services*

Note: Solid lines – flow of 'energy or matter' (i.e. physical relationships); dotted lines – flow of information; shapes indicate different steps in the assessment process and highlight the different nature of each step

Source: adapted from de Groot, 2006

definition by stating in the synthesis report that: 'ecosystem services are the benefits people obtain from ecosystems' (MEA, 2005).

Based on the MEA (2005) and de Groot et al (2002), ecosystem functions and services are grouped here into four primary categories:

- *Production functions (or provisioning services)* consist of the processes that combine and change organic and inorganic substances through primary and secondary production into goods that can be directly used by mankind.
- *Regulation functions (or regulating services)* relate to the capacity of natural and semi-natural ecosystems to regulate essential ecological processes and life support systems through biogeochemical cycles and other biosphere processes. In addition to maintaining ecosystem (and biosphere)

health, they provide many services with direct and indirect benefits to humans such as clean air, water and soil, nutrient regulation, disturbance prevention, biological control and pollination.

- *Information functions (or cultural services)* are those services that contribute to human mental well-being. Major categories of cultural services associated with forests are aesthetic and recreational use, spiritual and religious services and importance to cultural heritage.
- *Habitat functions (or supporting services)* relate to the importance of ecosystems to provide habitat for various stages in the life cycles of wild plants and animals, which, in turn, maintain biological and genetic diversity and evolutionary processes. Since these species and their role in the global ecosystem maintain most of the other ecosystem functions and services, the maintenance of healthy habitats is a necessary requirement for the provision of all ecosystem goods and services, directly or indirectly.

Table 2.1 gives an overview of the main goods and services forests provide (both native and planted) based on a synthesis of the classification methods mentioned above.

Goods and services of plantation forests

Plantation forests can be defined as 'Forest or other wooded land of introduced species and in some cases native species, established through planting or seeding' (FAO, 2006) (see also Chapter 1). The importance of forest plantations is stated in the 'Forestry Principles' adopted by the United Nations Conference on Environment and Development (UNCED) in Brazil 1992, and reiterates the importance of intensively managed forest plantations. Principle 6d says: 'The role of cultivated forests ... as sustainable and environmentally sound sources of renewable energy and industrial raw materials should be recognized, enhanced and promoted. Their contribution to the maintenance of ecological processes, to offsetting pressure on primary/old growth forests, and to providing regional employment and development ... should be recognized and enhanced' (Shelton et al, 2001).

The functions and services of plantation forests are diverse. FAO for instance, makes a distinction between 'productive' and 'protective' plantations (FAO, 2006). Productive plantations are focused primarily on the production of industrial wood, fuelwood and non-wood forest goods (e.g. animal fodder, apiculture, essential oils, tan bark, cork, latex, food), whereas protective plantations are established to provide conservation, recreation, carbon sequestration, water quality control, erosion control and rehabilitation of degraded lands, which also includes landscape and amenity enhancement (e.g. Fuhrer, 2000; Shelton et al, 2001; Lamb et al, 2005).

In Table 2.1, an overview is given of the main goods and services that can be provided by forests. A detailed description of all these goods and services is beyond the purpose of this chapter, but an attempt is made here to indicate, in our view, the differences in service provision between native and plantation forests based on an analysis of several publications. As shown in the table, the main service of plantations is the provision of resources (especially raw materials such as timber, energy sources such as palm oil, and to a lesser extent food, fodder and fertilizer). The reduction of surface run-off, erosion, and storm and flood damage are also important services. Compared to natural forests, however, the provision of most other forest services is reduced.

We stress that this is a preliminary analysis of limited literature sources, and much uncertainty still exists about how to measure service performance and the influence of management on the provision of goods and services. For example, it is generally assumed that forests prevent erosion and reduce run-off (e.g. Brown et al, 2005), but exactly to what extent, and whether this varies between different forest types is not well understood (e.g. Bruijnzeel, 2004).

Quantification and valuation of forest goods and services

Identification and quantification

In order to assess the capacity of forests (native or planted) to provide goods and services, indicators are needed. For each of the four categories of goods and services described above (provisioning, regulating, cultural and supporting), the main ecological processes associated with the service provision, and the main indicators determining the capacity of the system to provide the service are listed in Table 2.1. These indicators can be measured in different ways (e.g. literature review, expert consultations, community meetings, field visits or experiments).

One problem with determining sustainable use levels is that many goods and services depend on the same function (i.e. second column in Table 2.1), which means that use of one good or service will influence the availability of another. For example, a continuous supply of timber will depend directly on production functions (biomass production), which, in itself, will depend on habitat functions (i.e. suitable conditions for timber-producing species) and regulating functions (e.g. soil and climate regulation, pollination, etc.). Maintenance of these habitat and regulating functions will contribute, in turn, to the provision of other services (Campos et al, 2005).

In plantations, one service is usually maximized (e.g. timber production) at the expense of most other services and much external input (labour, energy, nutrients) is needed to maintain the productivity. The extent to which this trade-off is acceptable (ecologically, socio-culturally and economically) is the subject of the next section on 'valuation'.

Valuation

Once the capacity of an ecosystem to provide goods and services is known, their importance or 'value' can be determined. This importance or value primarily consists of three types of values: ecological, socio-cultural and economic (MEA, 2005).

Table 2.1 *Overview of forest services and examples of indicators for measuring function performance*

Services (comments and examples)	Ecological process and/or component providing the service (or influencing its availability) = Functions	State indicator (how much of the service is present)	Performance indicator (how much can be used /provided in sustainable way)
Provisioning			
Food	Presence of edible plants and animals	Total or average stock in kg/ha	
Fibre, fuel and other raw materials	Presence of species or abiotic components with potential use for timber, fuel or raw material	Total biomass (kg/ha)	
Biochemical products and medicinal resources	Presence of species or abiotic components with potentially useful chemicals and/or medicinal use	Total amount of useful substances that can be extracted (kg/ha)	Actual use in most appropriate unit per ha per year (in relation to maximum sustainable use level)
Genetic materials: genes for resistance to plant pathogens	Presence of species with (potentially) useful genetic material	Total 'gene bank' value (e.g. number of species and sub-species)	
Ornamental species and/or resources	Presence of species or abiotic resources with ornamental use	Total biomass (kg/ha)	
Regulating			
Air quality regulation: e.g. capturing dust particles	Capacity of ecosystems to extract aerosols and chemicals from the atmosphere	– Leaf area index – NOx-fixation, etc.	– Amount of pollutants 'extracted' – Effect on air quality
Climate regulation	Influence of ecosystems on local and global climate through land-cover and biologically mediated processes	– Greenhouse gas balance (esp. C-sequestration) – Land cover characteristics, etc.	Quantity of greenhouse gases etc. fixed and/or emitted $\rightarrow$ effect on climate parameters
Water quality regulation	Role of biota and abiotic processes in removal or breakdown of excess amounts organic matter, nutrients and polluting compounds	– Denitrification (kg N/ha/y) – Immobilization in plants and soil	Maximum amount of chemicals that can be recycled or immobilized on a sustainable basis
Water regulation	Role of forests in water infiltration and gradual release of water	Water retention capacity in soils, etc. or at the surface	Quantity of water retention and influence of hydrological regime (e.g. irrigation)

Table 2.1 *continued*

Services (comments and examples)	Ecological process and/or component providing the service (or influencing its availability) = Functions	State indicator (how much of the service is present)	Performance indicator (how much can be used /provided in sustainable way)
Erosion protection	Role of vegetation and biota in soil retention	– Vegetation cover – Root matrix	Amount of soil retained or sediment captured
Natural hazard mitigation	Role of forests in dampening extreme events (e.g. protection against flood damage)	Water-storage (buffer) capacity in m³	Reduction of flood-danger and prevented damage to infrastructure
Biological regulation	Control of pest populations through trophic relations; role of biota in distribution, abundance and effectiveness of pollinators	– Number and impact of pest-control species – Number and impact of pollinating species	– Reduction of human diseases, live-stock pests, etc – Dependence of crops on natural pollination
Cultural and Amenity			
Aesthetic: appreciation of natural scenery (other than through deliberate recreational activities)	Aesthetic quality of the landscape, based on e.g. structural diversity, 'greenness', tranquility	Presence of landscape features with stated appreciation	Expressed aesthetic value, e.g.: – No. of houses bordering natural areas – No. of users of 'scenic routes'
Recreational: opportunities for tourism and recreational activities	– Landscape features – Attractive wildlife	Presence of landscape and wildlife features with stated recreational value	– Maximum sustainable number of people and facilities – Actual use
Cultural heritage and identity: sense of place and belonging	Culturally important landscape features or species	Presence of culturally important landscape features or species	Number of people 'using' forests for cultural heritage and identity
Spiritual and artistic inspiration: nature as a source of inspiration for art and religion	Landscape features or species with inspirational value to human arts and religious expressions	Presence of landscape features or species with inspirational value	– No. of people who attach religious significance to forests, – No. of books, paintings, etc. using ecosystems as inspiration

Table 2.1 *continued*

Services (comments and examples)	Ecological process and/or component providing the service (or influencing its availability) = Functions	State indicator (how much of the service is present)	Performance indicator (how much can be used /provided in sustainable way)
Education and science opportunities for formal and informal education and training	Features with special educational and scientific value/interest	Presence of features with special educational and scientific value/interest	Number of classes visiting Number of scientific studies etc
Supporting			
Habitat and nursery service	Importance of ecosystems to provide breeding, feeding or resting habitat to resident or migratory species (and thus maintain a certain ecological balance and evolutionary processes)	– No. of resident, endemic species – No. of transient species – Habitat integrity – Minimum critical surface area of specific habitat	– 'Ecological Value' (i.e. difference between actual and potential biodiversity value) – Dependence of transient species on specific area

Ecological valuation

The 'Ecological Value', or importance of a given ecosystem, is determined mainly by the degree to which the ecosystem provides Regulation and Habitat Services, which, in turn, is measured by ecosystem criteria such as naturalness, diversity and rarity (see Table 2.2).

Whether plantation forests lead to a decrease in the overall ecological value of an area or landscape depends largely on the condition of the original ecosystem or production system they replaced. For example, ecological values are likely to increase if plantations are established on former agricultural land, and likely to decrease if established on land converted through the clearing of native ecosystems (see also Chapter 5).

Socio-cultural valuation

Social values (such as cultural diversity, identity, heritage and spiritual values) and perceptions play an important role in determining the importance of ecosystems and their services to human society (see Table 2.3). Native forests have many such values and are often an important source of non-material well-being to many individuals and societies.

Usually, socio-cultural values are reduced or lost when a native forest is replaced by a plantation. However, there are large differences depending on the

Table 2.2 *Ecological valuation criteria and measurement indicators*

Criteria	Short description	Measurement units/indicators
Naturalness/Integrity (representativeness)	Degree of human presence in terms of physical, chemical or biological disturbance.	– Quality of air, water and soil – % key species present – % of min. critical ecosystem size
Diversity	Variety of life in all its forms, including ecosystems, species and genetic diversity.	– number of ecosystems per geographical unit – number of species/surface area
Uniqueness/rarity	Local, national or global rarity of ecosystems and species	– number of endemic species and sub-species
Fragility/vulnerability (resistance)	Sensitivity of ecosystems to human disturbance	– energy budget (GPP/NPP) [1] – carrying capacity – complexity and diversity – succession stage/time
Resilience Renewability/ restorability	The possibility of spontaneous renewal or human aided restoration of ecosystems	

[1] GPP – Gross Primary Production; NPP = Net Primary Production

Source: de Groot et al, 2006

management system, and so-called 'community forests' (which are usually more or less heavily managed native forests), can have considerable socio-cultural importance.

To some extent these values can be captured by economic valuation methods (see further below), but these techniques do not capture fully the extent to which some ecosystem services are essential to people's very identity and existence.

Economic valuation

Finally, ecosystem services have economic importance, although some authors regard cultural values and their social welfare indicators as a sub-set of economic values. Others state that, in practice, economic valuation is limited to efficiency and cost-effectiveness analyses, usually measured in monetary units, and disregards the importance of, for example, spiritual values and cultural identity.

Therefore, in this chapter, economic and monetary valuation are treated separately from socio-cultural valuation. However, it is emphasized that ecological, socio-cultural and economic values all have their separate role in decision-making and should be seen as essentially complementary pieces of information in the decision-making process.

To analyse the economic value of ecosystems, the concept of Total Economic Value (TEV) (Figure 2.1) has become a framework widely used for quantifying the utilitarian value of ecosystems (de Groot et al, 2006). This framework typically disaggregates TEV into two categories: *use values* and *non-use values*.

Table 2.3 *Socio-cultural valuation criteria and measurement indicators*

Criteria	Short description	Measurement units/indicators
Therapeutic value	The provision of clean air, water and soil, space for recreation and outdoor sports and general therapeutic effects of nature on people's mental and physical well-being	– Suitability and capacity of natural systems to provide 'health services' – Restorative and regenerative effects on peoples' performance – Socio-economic benefits from reduced health costs and conditions
Amenity value	Importance of nature for cognitive development, mental relaxation artistic inspiration, aesthetic enjoyment and recreational benefits	– Aesthetic quality of landscapes – Recreational features and use – Artistic features and use – Preference studies
Heritage value	Importance of nature as reference to personal or collective history and cultural identity	– Historic sites, features and artefacts – Designated cultural landscapes – Cultural traditions and knowledge
Spiritual value	Importance of nature in symbols and elements with sacred, religious and spiritual significance	– Presence of sacred sites or features – Role of ecosystems and/or species in religious ceremonies and sacred texts
Existence value	Importance people attach to nature for ethical reasons (intrinsic value) and inter-generational equity (bequest value). Also referred to as 'warm glow-value'	– Expressed (through, for example, donations and voluntary work) or stated preference for nature protection for ethical reasons

Source: de Groot et al, 2006

Use values comprise three elements: direct use, indirect use and option values. *Direct use value* is also known as the extractive, consumptive or structural use value and derives mainly from *goods* that can be extracted, consumed or enjoyed directly (Dixon and Pagiola, 1998). *Indirect use* value is also known as the non-extractive use value, or functional value and derives mainly from the *services* the environment provides (Dixon and Pagiola, 1998). *Option value* is the value attached to maintaining the option to take advantage of the use value of something at a later date. Some authors also distinguish *quasi option value*, which derives from the possibility that even though something appears unimportant now, information received later might lead us to re-evaluate it (Dixon and Pagiola, 1998).

Non-use values, as the name says, derives from benefits the environment may provide when it is not used in any way. In many cases, the most important benefit of this kind is *existence value*; the value people derive from the knowledge that something exists even if they never plan to use it. Thus people place a value on the existence of blue whales or the panda even though they have never seen one and probably never will. However, if blue whales became

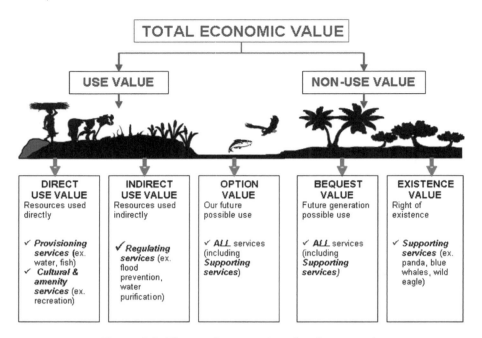

Figure 2.2 *The total economic value framework*

Note: 'bequest value' is often also shown as another kind of (future) use (option) value.
Source: adapted from MEA (2003), based on Pearce and Warford (1993) and Dixon and Pagiola (1998)

extinct, many people would feel a definite sense of loss (Dixon and Pagiola, 1998). *Bequest value* is the value derived from the desire to pass on values to future generations (i.e. our children and grandchildren).

Monetary value of ecosystem goods and services

Many of the above-mentioned values can be quantified in monetary terms. Monetary or financial valuation methods fall into three basic types, each with its own repertoire of associated measurement issues (Table 2.4); direct market valuation, indirect market valuation and survey-based valuation (i.e. contingent valuation and group valuation).

If no site-specific data can be obtained (due to lack of data, resources or time), then benefit transfer can be applied (i.e. by using results from other, similar areas to approximate the value of a given service in the study site). This method is rather problematic because, strictly speaking, each decision-making situation is unique. However, as more data become available from new case studies, benefit transfer becomes more reliable.

Although Table 2.4 is based on various literature sources, and seeks to reflect a broad consensus on monetary valuation methods, other views and terminologies do exist. For example, Dixon and Pagiola (1998) use the term

Table 2.4 *Monetary valuation methods, constraints and examples*

Method		Description	Constraints	Examples
1. Direct Market Valuation	Market price	The exchange value (based on marginal productivity cost) that ecosystem services have in trade	Market imperfections and policy failures distort market prices	Mainly applicable to the 'goods' (e.g. timber) but also some cultural (e.g. recreation) and regulating services (e.g. pollination)
	Factor income or productivity factor method	Measures effect of ecosystem services on loss (or gains) in earnings and/or productivity)	Care needs to be taken not to double-count values	Reversal of soil degradation which increases site quality and ecosystem productivity and thereby incomes of forest owners
	Public pricing *	Public investments, e.g. land purchase, or monetary incentives (taxes/subsidies) for ecosystem service use or conservation	Property rights sometimes difficult to establish; care must be taken to avoid perverse incentives	Investments in watershed protection to provide drinking water, or conservation measures
2. Indirect Market Valuation	Avoided (damage) cost method	Services that allow society to avoid costs that would have been incurred in the absence of those services	It is assumed that the costs of avoided damage or substitutes match the original benefit; however, this match may not be accurate, which can lead to underestimates as well as overestimates.	The value of the flood control service can be derived from the estimated damage if flooding would occur
	Replacement cost and substitution cost	Some services could be replaced with human-made systems		The value of groundwater recharge can be estimated from the costs of obtaining water from another source (substitute costs)
	Mitigation or restoration cost	Cost of moderating effects of lost functions (or of their restoration)		E.g. cost of preventive expenditures in absence of wetland service (e.g. flood barriers) or relocation
	Travel cost method	Use of ecosystem services may require travel and the associated costs can be seen as a reflection of the implied value	Overestimates are easily made; the technique is data intensive	E.g. part of the recreational value of a site is reflected in the amount of time and money that people spend while travelling to the site
	Hedonic pricing method	Reflection of service demand in the prices people pay for associated marketed goods	The method only captures people's willingness to pay for perceived benefits; very data intensive	For example: clean air, presence of water and aesthetic views will increase the price of surrounding real estate

Table 2.4 *continued*

Method		Description	Constraints	Examples
3. Surveys	Contingent valuation method (CVM)	This method asks people how much they would be willing to pay (or accept as compensation) for specific services through questionnaires or interviews	There are various sources of bias in the interview techniques; also there is controversy over whether people would actually pay the amounts they state in the interviews	It is often the only way to estimate non-use values; for example, a survey questionnaire might ask respondents to express their willingness to increase the level of water quality in a stream, lake or river so that they might enjoy activities like swimming, boating, or fishing
	Group valuation	Same as contingent valuation (CV) but then as an interactive group process	The bias in a group CV is supposed to be less than in individual CV	
4. Benefit Transfer		Uses results from other, similar areas, to estimate the value of a given service in the study site	Values are site and context dependent and therefore in principle not transferable	When time to carry out original research is scarce and/or data is unavailable, benefit transfers can be use (but with caution)

Source: In de Groot et al, 2006, compiled after Barbier et al, 1997; King and Mazotta, 2001; Wilson and Carpenter, 1999; Stuip et al, 2002

'change in output of marketable goods' as a combined term for market price and factor income; and they combine avoided (damage) cost, replacement cost and mitigation cost into so-called 'cost-based approaches'.

Each method has its advantages and disadvantages. Yet, in order to reach an approximation of the true contribution of native forests and plantations to human welfare and the economy, a mix of these methods should be applied to include all functions, and associated goods and services, in the decision-making process.

In Table 2.5, a first attempt is made to quantify the difference in service performance between native and planted forests.

By using a mix of valuation methods, based on over 100 case studies, Costanza et al (1997) came to a conservative estimate that the total economic value of native forests is, on average, between US$300 and 2000/ha/year (1994 values).

Although there are very few data available about the monetary value of ecosystem services provided by plantations, the data that are available indicate that the (average) TEV of plantations is lower than the average TEV of native forests (Fisher et al, 2008).

Table 2.5 *Difference in provision of goods and services between natural and plantation forests*

Main ecosystem services categories	Goods and services provided by forests		Natural (2) US$/ha/ year (1)	Plan-tation (3)
Provisioning Services	**Food** (from harvesting forest wildlife or gathering plant-products)	3	43	0
	Raw materials (e.g. timber, fibre)	1	138	+
	Energy resources (e.g. fuelwood, biofuels)	2	(incl. in above)	+
	Fodder and fertilizer (e.g. leaves, other organic matter)	1	(incl. in above)	0
	Genetic resources (genes and genetic information used for animal and plant breeding and biotechnology)	2	16	–
	Natural medicines and pharmaceuticals (e.g. drugs, models, tools, essay org.)	2	(incl. in above)	–
	Biochemicals (non-medicinal) (e.g. for dyes, biocides, food-additives)	2	(incl. in above)	–
	Ornamental resources: wildlife used in e.g. fashion, handicraft, jewellery, worship, souvenirs, decoration, as pets and in landscaping	2	(incl. in raw materials)	–
Regulating Services (regulation functions)a	**Air quality regulation** (e.g. capturing dust particles, NOx fixation, etc)	2	87	–
	Climate regulation Including carbon sequestration and storage	2	141	–
	Water quality regulation (filtering of rainwater and run-off water)	2	3	–
	Water regulation (buffering of extremes in run-off and river discharge)	2	2	–
	Natural hazard regulation (reduction of storm and flood damage)	1	2	0
	Erosion prevention (soil retention and prevention of landslides/siltation) and **maintenance and restoration of productive soils**	3	96 10	–
	Biological control (reduction/prevention of crop, livestock and/or human diseases by providing a barrier or habitat for control of vectors)	1	2	– –
	Pollination (providing habitat for pollinators of crops and wild plants)	2	(incl. in above)	–

Table 2.5 *continued*

Main ecosystem services categories	Goods and services provided by forests		Natural (2)	Plantation (3)
Cultural and Amenity Services (information functions)	**Aesthetic information** (non-recreational enjoyment of scenery)	2	(incl. in recr.)	−
	Recreation and nature-based tourism	2	66	−
	Cultural heritage and identity (many people value a 'sense of place' which is often associated with forests)	2	2	−
	Inspiration (e.g. for art, folklore, national symbols, architecture, design, advertising	2	(incl. in cultural)	−
	Spiritual and religious information (many individuals and religions attach spiritual values to forests and/or individual species)	2	(incl. in cultural)	−
	Educational information (both formal and informal education in nature)	2	(incl. in cultural)	− −
	Science (ecosystems, incl. forests influence the type of knowledge systems developed by different cultures)	2	(incl. in cultural)	− −
Supporting Services (habitat functions)	**Refugium** (provide habitat for resident plants and animals and migratory species and thus contribute to maintenance of biodiversity and evolutionary processes	3		− − −
	Nursery (provide reproduction habitat for species with commercial value that spend their adult life elsewhere)	?		?

(1) Average monetary value of this service US$/ha/year (based on meta-analysis of over 100 studies by Costanza et al, 1997)

(2) Qualitative scale indicates relative performance of natural forests in providing the given services: 1 = low, 2 = medium, 3 = high

(3) +/0/− indicates difference in services-provision between natural and plantation forest (+ = service is enhanced, 0 = remains the same, − = service is reduced)

(*) based on publications mentioned in the table-title. Of course there are large differences between forest types and management regimes so this must be seen as a very rough indication and further study is needed.

Source: qualitative scale based on interpretation of information from Brown and Lugo, 1990; Parotta et al, 1997, 2002; Shelton et al, 2001; de Groot et al, 2002; MEA, 2003).

Yet native forests are still converted into single-function land-use types (e.g. croplands) on a large scale even though proof is mounting that the total value of multi-functional use of natural and semi-natural landscapes is, in the long term, often economically more beneficial than short-term economic benefits generated by the converted systems (Balmford et al, 2002). Clearly, these economic calculations must be interpreted with care but they can help to highlight the economic (and financial) implications of these ecosystem conversions and identify 'winners and losers', which can help to develop financing schemes for sustainable management. If all forest services were valued properly, and landholders received some of this value, then plantation establishment might become financially attractive in degraded areas where they could be environmentally very beneficial (e.g. to reduce erosion, improve control of the water table and provide resources and other services).

Payment for ecosystem services (PES) and carbon markets

There are increasing efforts to develop a payment mechanism for ecosystem services (PES), which can be regarded as another approximation of the monetary value of ecosystem goods and services (e.g. Wunder, 2005, 2007). Payments for ecosystem services may compensate for the imbalance between the (public) costs of avoiding deforestation and the (private) income generation from the conversion of (tropical) forests to other land use (e.g. Scherr et al, 2004). PES mechanisms have been designed for a wide range of services, including forest preservation, water retention and biodiversity conservation. For example, a PES programme could require participants to end all forest-damaging activities (e.g. by establishing a private reserve), or simply maintain a minimum amount of canopy cover while limiting road development in the reserve. Gene (2007) indicates that protection of a forest in Costa Rica became an attractive alternative against logging when the annual award of PES was set at around US$75/ha/yr.

Carbon financing schemes like the Clean Development Mechanism (CDM) and Reduced Emissions from Deforestation and Degradation (REDD) could help to secure both the economic and ecological stability of forested areas (e.g. Pfaff et al, 2000, 2007). Moreover voluntary CO_2 emissions trading may be a mechanism that will enhance the amount of money people receive for the protection and sustainable use of forested areas. These mechanisms are different in various ways, yet they have one thing in common: they only pay for the so-called additionality; the surplus of reduced CO_2 due to human activities.

Forest plantations have a relatively high wood production rate (and thus carbon storage) in comparison to many native forests (Evans and Turnbull, 2004). Therefore the Clean Development and emissions trading mechanisms are thought to be important new incentives for investments in forest plantations in the tropics (e.g. FAO, 2006). Jindal et al (2008) give an overview of African carbon projects, which include many plantation projects. They indicate that cash incomes for households can be increased significantly. In one project in Mozambique, local households received a cash payment of US$242 per ha over seven years for carbon sequestered on their farms. Other benefits to farmers included access to fruits, minor timber, firewood and any other non-timber forest products (Jindal et al, 2008). However, to date there is still much discussion about whether, and how, international mechanisms actually can provide financial support to developing countries interested in undertaking REDD and CDM programmes (e.g. CCAP, 2009).

Quantification and valuation of local goods and services provided by planted forests in Northern India: a case study

Introduction

In this section, results of a study of the relative contribution of planted forests in India to local livelihoods are briefly presented and discussed. The study was carried out as part of the NETFOP project (NETworking Forest Plantations in a crowded world: optimizing ecosystem services through improved planning and management strategies), an EU funded project under the India ECCP (Economic Cross-Cultural Programme). The project (2005–2006) aimed to establish an expert network on forest plantation planning and management in three densely populated countries (India, Germany and The Netherlands) (van der Meer et al, 2007a).

Forests in India cover some 67.7 million hectares (Mha), which is about 21 per cent of the total land area (FAO, 2006). According to FAO (2006), India has some 3.2Mha of forest plantations with some 1Mha in productive and 2.2Mha in protective plantations). Eucalypts (*Eucalyptus* spp.), acacia (*Acacia* spp.) and teak (*Tectona grandis*) are the main plantation species with a share of 25 per cent, 20 per cent and 8 per cent respectively. Between 2000 and 2005, the area of forest plantations increased by some 84,000ha/yr (FAO, 2006). About 60 per cent of the 1.1 billion Indian population is dependent on the forests for energy, livestock grazing and construction materials (FSI, 2003). As the Indian population is increasing rapidly, there is an urgent need for sound planning and sustainable management to prevent over-use and degradation of India's forest resource. In addition, recognition of the ability of planted forests to supply a wide range of goods and services that are important for the rural people of India has increased (e.g. Gundimeda et al, 2007).

The study was carried out in Uttaranchal (Uttarakhand), a state in the mountainous region of Northern India with a total land area of 53,483km² and a population of 8.5 million. Some 45.7 per cent of the state area is covered with forests (FSI, 2005). The major forest types are tropical moist deciduous forest, tropical dry deciduous forest, subtropical pine forest, Himalayan dry temperate forest and (sub)alpine forest. Almost 27 per cent of the state is covered by moderately dense forests (canopy cover 40–70 per cent), while open forests (canopy cover 10–40 per cent) cover some 11 per cent of the state. Approximately 7.5 per cent of the state is covered by very dense (>70 per cent cover) forests (FSI, 2005).

The area of Uttranchal covered by forest plantations, is difficult to estimate. As an approximation, the value for 'tree cover' may be used, which, in Uttaranchal, is 1.23 per cent (658km²) (FSI, 2005). This indicates the area of the state covered by small patches of trees (<1.0ha in area), which are not recorded as forest area. This area includes trees in small-scale plantations, woodlots or scattered trees on farms, homesteads and urban areas. Trees along roads and canals also are included. However, plantations larger than 1ha in size are not included, meaning that the area covered by forest plantations, which more appropriately may be called cultivated forest, is likely to occur on more than 1.23 per cent of the state land area.

Identification and quantification of ecosystem goods and services

To determine what products and services were used from the different forest areas, and what value they represented, a multidisciplinary landscape assessment (MLA) was performed. The MLA consists of a set of methods initially developed by CIFOR and partners to address biodiversity conservation in local communities in Kalimantan (Sheil et al, 2003).

In each of the three following altitudinal zones two villages were selected:

- the lower zone (600–1000m), vegetation dominated by sal (*Shorea robusta*) forests and associated species;
- the middle zone (1000–1600m), with forests dominated by chir pine (*Pinus roxburghii*);
- the upper zone (1600–2200m), with forests dominated by deodar (*Cedrus deodara*).

The size of the six villages varied between 250 and 900 inhabitants. The number of households per village varied between 35 and 103, with an average of 7 persons per household.

In each of the villages, a community meeting was organized to identify the landscape units surrounding the village, and what goods and services each of these landscape units delivered to the villagers. Subsequently more detailed information was gathered from selected households in each village by focus group interviews. The number of households interviewed per village ranged from 5 to 10. The information from the community meetings and household interviews was used to estimate the quantity of EGS used in each of the four groups of goods and services identified (provisioning, regulation, supporting and cultural). In addition, information on harvesting patterns and people's perceptions of intangible services was collected through questionnaires.

Ranking and valuation of EGS

A scoring exercise (pebble distribution method: PDM) was used to assess the importance of various landscape units for providing different ecosystem goods and services. This method is relatively easy to use, and is commonly employed to rank the importance of various landscape units for the provision of goods and services (e.g. Sheil et al, 2003, Sheil and Liswanti, 2006).

The importance value of the identified EGS was also determined in economic, ecological and social terms following the approach described above. However, as we aim to illustrate the application of EGS to, and the economic valuation of, planted forests in India, these results are not presented here. A complete overview of the EGS provided by Indian forests, and the value they represent is beyond the scope of this chapter.

The economic valuation focused on provisioning goods and services, as these are the main sources of income for local people living in and around the forested areas studied. The direct market value method was used to determine the total value of EGS produced by the forest landscapes.

Results
Identification of EGS

During community meetings, the following landscape units in and around villages were identified:

1 State forest: forest managed under the Forest Department; these include both planted and native forest.
2 Community forest: common property under the Revenue Department. Local statutory bodies (e.g. Van Panchayats) are responsible for management; plantations have been established on this land with support of the Watershed Department.
3 Grassland, community land.
4 Rivers and streams: usually in the valleys and lower parts.
5 Village area: land occupied by human settlements, including trees growing on this land.
6 Agricultural land: private land owned by households for growing crops.
7 Fallow land: all arable land not occupied by crops in order to regain its fertility.

The household survey produced a list consisting of four categories of goods and services (provisioning, regulating, supporting and cultural), each of which consists of six to ten goods and/or services:

- Provisioning EGS: (1) food, (2) raw materials (timber, fibre, baskets), (3) energy resources (fuelwood, cow dung), (4) cattle-related products (fodder, cattle bed leaves), (5) agriculture-related products (manure and others), (6) genetic resources, (7) medicinal resources, (8) ornamental resources, (9) cultivation (grazing), (10) waste treatment (sanitary facilities).
- Regulation EGS: (1) water flow regulation, (2) erosion prevention, (3) water quality maintenance, (4) soil quality maintenance, (5) natural hazard regulation, (6) air quality regulation, (7) climate regulation, (8) biological control, (9) pollination.
- Cultural EGS: (1) Aesthetic information, (2) recreation and nature-based tourism, (3) cultural heritage and identity, (4) inspiration, (5) spiritual and religious information, (6) educational information, (7) science.
- Supporting EGS: (1) refuge area, (2) nursery, (3) primary production, (4) nutrient cycling, (5) soil formation, (6) water cycling.

Quantification and valuation of provisioning EGS

For each of the ten categories of the provisioning EGS, the total annual harvest from all landscape-unit types surrounding the village are given in Table 2.6. In the third column, the percentage harvested from forest areas (state forests and community forest) is given. It shows that, overall, 64 per cent of all provisioning goods and services in the study area are provided by forests.

Table 2.6 *Quantities of annual harvest from forest landscapes (state and community forest)*

	Total quantity (kg/hh⁻¹/yr⁻¹)	% from forest
Food products	38	68%
Raw materials (timber, fibre, baskets)	77	69%
Energy (fuelwood)	1794	69%
Cattle-related products (fodder, cattle-bed)	6603	58%
Agriculture-related products (manure etc.)	8215	61%
Genetic resources[1]	3 (no/year)	90%
Medicinal resources	0.6	30%
Ornamental resources[2]	(17% of resp.)	60%
Cultivation (grazing)	16,419	67%
Waste treatment[3])	(62 % of resp.)	34%
TOTAL	33,146	64%

Quantities are given in kilograms harvested per household (hh) per year unless otherwise indicated. (average hh size = 7 persons).

There was considerable variation in the quantity of EGS used between the three altitudinal zones (see van der Meer, 2007a). In general, inhabitants from villages in the middle zone used the largest quantities of the various goods and services. Differences between the use of EGS by inhabitants from the lower and the upper zones were less conspicuous. For instance, in one of the middle zone villages, every household consumed 82kg of edible forest items per year, compared to 29kg and 4kg in the villages in the upper and lower zone, respectively.

The direct market value method was used to determine the total monetary value of EGS produced by the forest landscapes (Table 2.7). In the economic valuation, the ornamental, medicinal and genetic resources were not considered as the quantities reported were very small and varied considerably between households.

Table 2.7 *Monetary value (in US$) of annual benefits from forest landscapes (state and community forest)*

Provisioning services	Quantities from forest (kg·hh-1·yr-1)	Direct market value (US$·hh-1·yr-1)
Food products	26	17
Raw materials (timber, fibre, baskets)	53	18
Energy (Fuelwood)	1229	73
Cattle-related products (fodder, cattle-bed)	3815	68
Agriculture-related products (manure and others)	4997	37
Cultivation (grazing)	10,939	263
Waste treatment	(34% of respondents).	1
TOTAL		477

Source: van der Meer et al, 2007b

Tables 2.6 and 2.7 show that people in the selected villages enjoy a wide range of provisioning services from the forest. The total value of the forest services (US$477 per annum) is equal to a substantial part (around 80 per cent) of the average annual income in the area, which is about US$638/hh/yr.

Discussion

Forests, either planted or (semi-)natural, play an important role in the provision of EGS to local communities in Northern India as in many other parts of the (less-developed) world. Gundimeda et al (2007) indicate that, in particular, the income from forests in the north-eastern states of India is significantly underestimated by traditional measures of income. Other studies done in the same region confirm the important role of forests in the provision of various goods and services like fuelwood (e.g. Maikhuri, 1991; Shyamsundar and Kramer, 1997), non-timber forest products (Mahapatra and Tewari, 2005; Murthy et al, 2005) and medicinal plants (van de Kop et al, 2006).

The observed variation in quantity and value of utilized EGS between villages is probably determined not only by the differences in altitude and associated vegetation; villages in the lower zone were close to urban centres providing jobs and income, and villages in the higher zone obtained income from tourism. The villages in the middle zone were quite isolated with the majority of people depending most heavily on forests for their daily living supplies. The economically less-developed villages were more dependent on provisioning services than more-developed villages; e.g. the collection of cattle bed leaves and fodder was almost twice as high in the remote villages compared to the other villages (e.g. 3300 vs. 1800kg/annum/hh). Also the amount of fuelwood collected was higher in remote villages than in the other villages.

Here we only described the quantification and valuation of the provisioning goods and services. This does not mean that other regulating, supporting and cultural services were not important. Our surveys suggest that, according to local perception, planted forests were the most important landscape unit providing regulating (e.g. soil and water control) and information services (e.g. inspiration, education) (van der Meer et al, 2007a). Other studies have confirmed the important role of both planted and natural forests in providing water and soil conservation (e.g. Guo and Li, 2003; Aylward, 2004; Wang et al, 2009). However, the quantification and valuation of these less tangible services is often problematic. For instance, it is not easy to quantify the amount of water flow regulated or the amount of soil improved (see Chapter 4). Second, it is not easy to get actual payments for the indirect economic value of forest ecosystem services such as the regulation of hydrological cycles and soil conservation, as for instance indicated by Scherr et al (2004) and Wunder (2007).

Although it may be clear that forest (plantations) in some cases are able to sustain and enhance local biodiversity levels (e.g. Cossalter and Pye-Smith, 2003), it proved difficult to quantify biodiversity levels because this requires in-depth and often long-term investigations, which were beyond the scope of this study. Also translating biodiversity into a (monetary) value is difficult. Conservation and use of the services biodiversity provides requires financing on a large scale (e.g. Jenkins et al, 2004); payment for biodiversity on a local scale currently seems unrealistic.

The EGS methodology enables quantification and valuation of the goods and services of planted forests. The quantification and (monetary) valuation of provisioning goods and services from planted forests showed that rural communities, especially, in India depend largely on forests for their daily livelihood. The quantification of the regulating, supporting and cultural services proved to be more difficult. However, using a scoring methodology, it was possible to indicate the relative importance and rank the various goods and services in these categories.

The economic valuation of provisioning services is most straightforward, partly because these goods and services are relatively easy to quantify. However, the ecological and cultural evaluation of the provisioning goods needs to be developed further. The monetary valuation of the regulating, supporting and cultural services is problematic. This is partly due to the difficult quantification, but also because there is no market for these goods and services.

Concluding remarks

In general, understanding the synergies and conflicts between different land uses (and land-use policies) is crucial for developing an integrated landscape approach. Lamb et al (2005) stress that restoration of degraded forest lands in tropical regions should focus on establishing an optimal mix of biodiversity, regulation functions and supply of goods and services. A sound understanding of the quantity and value of the goods and services is much needed to improve planning and managing planted forests. This means that information is needed on both biophysical processes, like forest growth and water regulation, as well as on socio-economic factors, like population pressure and economic values. In acquiring this information it is important to involve all concerned stakeholders (e.g. Sayer et al, 2004; ITTO, 2005). Once there is a proper understanding of what forest functions are needed, it is possible to use spatial planning to design a forest landscape that delivers the optimal mix of goods and services (e.g. Sayer et al, 2004).

Planted forests afford opportunities to refine further spatial planning and the management of forest goods and services. For instance, the inclusion of high conservation value areas and biodiversity corridors could help to improve biodiversity levels in plantation areas without affecting the production function (e.g. Barlow et al, 2006; Cyranoski, 2007). Van Eupen et al, (2007) use scenario-dependent maps, which indicate habitat suitability of landscapes for certain flagship species. When integrated with planning for the other most important provisioning, regulating, cultural and supporting goods and services, this could be a valuable tool for better planning in planted forests. This could be used in the new approaches currently being developed for the restoration of degraded forest landscapes in tropical areas, which address both the sustainable use of biodiversity as well as alleviation of rural poverty (Lamb et al, 2005).

We conclude that a better understanding of the quantity and value of the goods and services provided by planted forests enables improved policy-making and management of (degraded) forested areas. Turning this into a 'real' money value (through PES schemes) can help to pay for the conservation, restoration and sustainable use of a forest area. These new policies and action plans are needed urgently to help restore a wide range of ecosystem services in the large area of degraded forest land in the tropics.

Acknowledgements

We express our gratitude to the European Union for providing the necessary funding under the EU–India Economic Cross Cultural Programme (EU-ECCP) to undertake the case study in India. We also express our thanks and gratitude to all the key persons, villagers, state forest and municipal officials, and all those who gave their cooperation and support in carrying out the surveys. Miriam van Heist is acknowledged for guiding the methodology for the field assessment in India. We thank Christiana Bairaktari, Febi Djafar, Hetty Mathijssen and Wondy Wondwossen for collecting the field data in India. Jürgen Bauhus, Anil Hooda, Joachim Schmerbeck and Madhu Verma are gratefully acknowledged for their valuable comments on the manuscript. The knowledge support programme of the Ministry of Food Quality, Nature and Agriculture, The Netherlands, supported this work through the KB1 – SELS project 2008.

References

Aylward, B. (2004) 'Land use, hydrological function and economic valuation', in Bonell, M. and Bruijnzeel, L. A. (eds) *Forests, Water and People in the Humid Tropics: Past, Present and Future Hydrological Research for Integrated Land and Water Management*, Cambridge University Press, Cambridge, pp99–120

Balmford, A., Bruner, A., Cooper, P., Costanza, R., Farber, S., Green, R. E., Jenkins, M., Jefferiss, P., Jessamy, V., Madden, J., Munro, K., Myers, N., Naeem, S., Paavola, J., Rayment, M., Rosendo, S., Roughgarden, J., Trumper, K. and Turner, R. K. (2002) 'Ecology: economic reasons for conserving wild nature', *Science*, vol 297, no 5583, pp950–953

Barlow, J., Peres, C. A., Henriques, L. M. P., Stouffer, P. C. and Wunderle, J. M. (2006) 'The responses of understorey birds to forest fragmentation, logging and wildfires: an Amazonian synthesis', *Biological Conservation*, vol 128, no 2, pp182–192

Brown, A. E., Zhang, L., McMahon, T. A., Western, A. W. and Vertessy, R. A. (2005) 'A review of paired catchment studies for determining changes in water yield resulting from alterations in vegetation', *Journal of Hydrology*, vol 310, pp28–61

Brown, S. and Lugo, A. E. (1990) 'Tropical secondary forests', *Journal of Tropical Ecology*, vol 6, pp1–32

Bruijnzeel, L. A. (2004) 'Hydrological functions of tropical forests: not seeing the soil for the trees?' *Agriculture, Ecosystems and Environment*, vol 104, no 1, pp185–228

Campos, J. J., Alpízar, F., Louman, B., Parrotta, J. and Porras, I. (2005) 'An integrated approach to forest ecosystem services', in Mery, G., Alfaro, R., Kanninen, M. and Lobovikov, M. (eds) *Forest in the Global Balance – Changing Paradigms*, IUFRO World Series, vol 17, pp97–116

Carter, J. (1996) *Participatory Forest Resource Assessment Methods: Recent Approaches to Participatory Forest Resource Assessment. Rural Development Forestry, Study Guide 2*, Rural Development Network, Overseas Development Institute, London

CCAP (2009) *Utilizing Payments for Environmental Services for Reducing Emissions from Deforestation and Forest Degradation (REDD) in Developing Countries: Challenges and Policy Options*, Center for Clean Air Policy, Washington, DC

Cossalter, C. and Pye-Smith, C. (2003) 'Fast-wood forestry: myths and realities', *Forest Perspectives*, CIFOR, Bogor, Indonesia

Costanza, R. and Farber, S. (2002) 'Introduction to the special issue on the dynamics and value of ecosystem services: integrating economic and ecological perspectives', *Ecological Economics*, vol 41, pp367–373

Costanza, R., Arge, R. D., Groot, R. D., Farber, S., Grasso, M., Hannon, B., Limburg, K., Naeem, S., O'Neill, R. V., Paruelo, J., Raskin, R. G., Sutton, P. and Belt, M. V. D. (1997) 'The value of the world's ecosystem services and natural capital', *Nature*, vol 387, pp253–260

Cyranoski, D. (2007) 'Logging: the new conservation', *Nature*, vol 446, no 7136, pp608–610

Daily, G. C. (1997) 'Ecosystem services: benefits supplied to human society by natural ecosystems', *Issues in Ecology 2*, Ecological Society of America, Washington, DC

de Groot, R. S. (1992) *Functions of Nature: Evaluation of Nature in Environmental Planning, Management and Decision-making*, Wolters Noordhoff BV, Groningen, The Netherlands

de Groot, R. S. (2006) 'Function analysis and valuation as a tool to assess land use conflicts in planning for sustainable, multi-functional landscapes', *Landscape and Urban Planning*, vol 75, pp175–186 de Groot, R. S., Wilson, M. and Boumans, R. (2002) 'A typology for the description, classification and valuation of ecosystem functions, goods and services', in *The Dynamics and Value of Ecosystem Services: Integrating Economic and Ecological Perspectives*, special issue of *Ecological Economics*, vol 41, no 3, pp393–408

de Groot, R. S., Stuip, M., Finlayson, M. and Davidson, N. (2006) *Valuing Wetlands: Guidance for Valuing the Benefits Derived from Wetland Ecosystem Services*, Ramsar Technical Report, no 3, CBD Technical Series, no 27

Dixon, J. and Pagiola, S. (1998) *Economic Analysis and Environmental Assessment*, Environmental Assessment Sourcebook Update, no 23, The World Bank

Evans, J. and Turnbull, J. W. (2004) *Plantation Forestry in the Tropics: The Role, Silviculture, and Use of Planted Forests for Industrial, Social, Environmental, and Agroforestry Purposes*, Oxford University Press, Oxford, XIII

FAO (2005) *Global Forest Resources Assessment 2005*, Food and Agriculture Organization, Rome

FAO (2006) 'Responsible management of planted forests: voluntary guidelines', Planted Forests and Trees Working Paper No. 37/E, Rome

Fisher, B., Turner, K., Zylstra, M., Brouwer, R., Groot, R. d., Farber, S., Ferraro, P., Green, R., Hadley, D., Harlow, J., Jefferiss, P., Kirkby, C., Morling, P., Mowatt, S., Naidoo, R., Paavola, J., Strassburg, B., Yu, D. and Balmford, A. (2008) 'Ecosystem services and economic theory: integration for policy-relevant research', *Ecological Applications*, vol 18, no 8, pp2050–2067

FSI (Forest Survey of India was SFR) (2003) *State of Forest Report 2003*, Forest Survey of India. Ministry of Environment and Forests, Dehradun, India

FSI (2005) *State of Forest Report, 2005*, Forest Survey of India, Ministry of Environment and Forests, Dehradun, India

Fuhrer, E. (2000) 'Forest functions, ecosystem stability and management', *Forest Ecology and Management*, vol 132, pp29–38

Gene, E. I. (2007) 'The profitability of forest protection versus logging and the role of payments for environmental services (PES) in the Reserva Forestal Golfo Dulce, Costa Rica', *Forest Policy and Economics*, vol 10, no 1–2, pp7–13

Gundimeda, H., Sukhdev, P., Sinha, R. K. and Sanyal, S. (2007) 'Natural resource accounting for Indian states: illustrating the case of forest resources', *Ecological Economics*, vol 61, pp635–649

Guo, Z., Gan, Y. and Li, Y. (2003) 'Spatial pattern of ecosystem function and ecosystem conservation', *Environmental Management*, vol 32, no 6, pp682–692

IUCN (2007) 'Forest landscape restoration: see the bigger picture. Global partnership on forest landscape restoration', International Union for Conservation of Nature, www.iucn.org/themes/fcp/publications/brochures_papers.htm, accessed 18 February 2008

ITTO (2005) 'Restoring forest landscapes: an introduction to the art and science of forest landscape restoration', International Tropical Timber Organization/IUCN, ITTO Technical Series no 23

Jenkins, M., Scherr, S. J. and Inbar, M. (2004) 'Markets for biodiversity services: potential roles and challenges', *Environment*, vol 46, pp32–42

Jindal, R., Swallow, B. and Kerr, J. (2008) 'Forestry-based carbon sequestration projects in Africa: potential benefits and challenges', *Natural Resources Forum*, vol 32, pp116–130

Lamb, D., Erskine, P. D. and Parrotta, J. A. (2005) 'Restoration of degraded tropical forest landscapes', *Science*, vol 310, no 5754, pp1628–1632

Limburg, K. E., O'Neil, R. V., Costanza, R. and Farber, S. (2002) 'Complex systems and valuation', *Ecological Economics*, vol 41, no 3, pp409–420

Mahapatra, A. K. and Tewari, D. D. (2005) 'Importance of non-timber forest products in the economic valuation of dry deciduous forests of India', *Forest Policy and Economics*, vol 7, pp455–467

Maikhuri, R. K. (1991) 'Nutritional value of some lesser-known wild food plants and their role in tribal nutrition: a case study from northeast India', *Tropical Science*, vol 31, pp397–405

Maikhuri, R. K., Senwal, R. L., Rao, K. S. and Saxena, K. G. (1997) 'Rehabilitation of degraded community lands for sustainable development in Himalaya: a case study in Garhwal Himalaya, India', *International Journal of Sustainable Development and World Ecology*, vol 4, pp192–203

MEA (Millennium Ecosystem Assessment) (2003) *Ecosystems and Human Well-being: A Report of the Conceptual Framework Working Group of the Millennium Ecosystem Assessment*, World Resources Institute, Island Press, Washington, DC

MEA (2005) *Ecosystems and Human Well-being: Synthesis*, World Resources Institute, Island Press, Washington, DC

Murthy, I. K., Bhat, P. R., Ravindranath, N. H. and Sukumar, R. (2005) 'Financial valuation of non-timber forest product flows in Uttara Kannada district, Western Ghats, Karnataka', *Current Science*, vol 88, pp1573–1579

Parrotta, J. A., Turnbull, J. W. and Jones, N. (1997) 'Introduction: catalyzing native forest regeneration on degraded tropical lands', *Forest Ecology and Management*, vol 99, no 1–2, pp1–7

Parrotta, J. A., Francis, J. K. and Knowles, O. H. (2002) 'Harvesting intensity affects forest structure and composition in an upland Amazonian forest', *Forest Ecology and Management*, vol 169, no 3, pp243–255

Pearce, D. and Warford, J. (1993) *World Without End: economics, Environment and Sustainable Development*, Oxford University Press, Oxford.

Pfaff, A. S. P., Kerr, S., Hughes, R. F., Liu, S. G., Sanchez-Azofeifa, G. A., Schimel, D., Tosi, J. and Watson, V. (2000) 'The Kyoto protocol and payments for tropical forest: an interdisciplinary method for estimating carbon-offset supply and increasing the feasibility of a carbon market under the CDM', *Ecological Economics*, vol 35, no 2, pp203–221

Pfaff, A., Kerr, S., Lipper, L., Cavatassi, R., Davis, B., Hendy, J. and Sanchez-Azofeifa, G. A. (2007) 'Will buying tropical forest carbon benefit the poor? Evidence from Costa Rica', *Land Use Policy*, vol 24, no 3, pp600–610

Sayer, J., Chokkalingham, U. and Poulsen, J. (2004) 'The restoration of forest biodiversity and ecological values', *Forest Ecology and Management*, vol 201, no 1, pp3–11

Scherr, S. J., White, A. and Khare, A. (2004) 'For services rendered: the current status and future potential of markets for the ecosystem services provided by tropical forests', International Tropical Timber Organization, ITTO Technical Series 21

Sheil, D. and Liswanti, N. (2006) 'Scoring the importance of tropical forest landscapes with local people: patterns and insights' *Environmental Management*, vol 38, pp126–136

Sheil, D., Puri, R., Basuki, I., van Heist, M., Wan, M., Liswanti, N., Rukmiyati, Sardjono, M. A., Samsoedin, I., Sidiyasa, K., Chrisandini, Permana, E., Angi, E. M., Gatzweiler, F., Johnson, B., and Wijaya, A. (2003) *Exploring Biological Diversity, Environment and Local Peoples Perspectives in Forest Landscapes*, 2nd edn, Center for International Forestry Research, Ministry of Forestry and International Tropical Timber Organization, Bogor, Indonesia

Shelton, D., Cork, S., Binning, C., Parry, R., Hairsine, P., Vertessy, R. and Stauffacher, M. (2001) 'Application of an ecosystem services inventory approach to the

Goulburn broken catchment', in Rutherford, I., Sheldon, F., Brierley, G. and Kenyon, C. (eds) *Proceedings of the Third Australian Stream Management Conference August 27–29, 2001*, Cooperative Research Centre for Catchment Hydrology, Brisbane, pp157–162

Shyamsundar, P. and Kramer, R. (1997) 'Biodiversity conservation – At what cost? A study of households in the vicinity of Madagascar's Mantadia National Park', *Ambio*, vol 26, no 3, pp180–184

van de Kop, P., Alam, G. and Piters, B. D. S. (2006) 'Developing a sustainable medicinal plant chain in India, linking, people, markets and values', in Ruben, R., Slingerland, M. and Nijhoff, H. (eds) *Agro-food Chains and Networks for Development*. Springer, pp191–202

van der Meer, P. J., Schmerbeck, J., Hooda, A., Bairaktari, C., Bisht, N. S., Gusain, M. S., Singh, C. J., Naithani, A., Saxena, A., Olsthoorn, A. F. M. and de Groot, R. S. (2007a) *Assessment of Ecosystem Goods and Services of Forests in India, The Netherlands, and Germany*, NETFOP report no 1 (NETworking FOrest Plantations in a crowded world: optimizing ecosystem services through improved planning and management)

van der Meer, P. J., Bairaktari, C., de Groot, R. S., van Heist, M. and Schmerbeck, J. (2007b) *Methodology for Assessing Benefits of Afforested Community and Forest Land*, NETFOP report no 2 (NETworking FOrest Plantations in a crowded world: optimizing ecosystem services through improved planning and management)

van Eupen, M., Sedze Puchol, T., Sharma, S. D. and Vijayanand (2007) *Modelling the Distribution of Goods and Services at the Landscape Level*, NETFOP report no 4 (NETworking FOrest Plantations in a crowded world: optimizing ecosystem services through improved planning and management)

Wang, C. Y., van der Meer, P. J., Peng, M. C., Douven, W., Hessel, R. and Dang, C. L. (2009) 'Ecosystem services assessment of two watersheds of Lancang River in Yunnan, China with a decision tree approach', *Ambio*, vol 38, no 1, pp47–54

Wilson, M. A. and Howarth, R. B. (2002) 'Discourse-based valuation of ecosystem services: establishing fair outcomes through group deliberation', *Ecological Economics*, vol 41, no 3, pp431–443

Wunder, S. (2005) *Payments for Environmental Services: Some Nuts and Bolts*, CIFOR Occasional Paper No. 42, Center for International Forestry Research, Jakarta

Wunder, S. (2007) 'The efficiency of payments for environmental services in tropical conservation', *Conservation Biology*, vol 21, pp48–58

3
Managing forest plantations for carbon sequestration today and in the future

Hannes Böttcher and Marcus Lindner

Introduction

Despite the past and recent losses of extensive forest areas globally, the remaining forests turn over approximately one-twelfth of the atmospheric stock of carbon dioxide through gross primary production (GPP) every year (Malhi et al, 2002). Yet less than 1 per cent of the carbon turned over ends up in a long-term terrestrial carbon sink. Forest ecosystems contain various carbon pools with different turnover rates. The average residence time of carbon in forest biomass in unmanaged tropical systems ranges from 50 to 100 years (Vieira et al, 2005). In systems where carbon is released earlier through harvesting or natural disturbance, the average age of carbon is lower.

Currently, forest ecosystems in the Northern Hemisphere form a sink for atmospheric CO_2. The Intergovernmental Panel on Climate Change (IPCC) estimated the global net uptake of carbon by terrestrial ecosystems from 1990 to 1998 to be about 2.4Gt C per year (Bolin and Sukumar, 2000; Schimel et al, 2001). Several processes contributed to carbon sequestration in the sink. In addition to the positive influences of anthropogenic nitrogen deposition, increasing global temperatures and rising CO_2 (Schulze, 2006; Canadell et al, 2007a; Magnani et al, 2007), management changes are of particular importance (Houghton, 2003).

Today 89 per cent of forests in industrialized countries and countries in transition, and about 12 per cent of the total forest area in all developing countries is either managed according to a formal or informal management plan or has been designated as a conservation area (Wilkie et al, 2003). The area clearly is dependent on the definition of forest management adopted. Forest management, in general, can be referred to as the application of biological, physical, quantitative, social and policy principles to the regeneration, tending, utilization and conservation of forests to meet specified goals (Sampson and Scholes, 2000). Forest management, as defined here, comprises afforestation and the implementation of activities such as planting, thinning and harvesting applied in existing forests.

In view of the properties of carbon stocks in forest ecosystems and the forestry sector, and the definition of forest management above, three known strategies are available to curb the increase of CO_2 in the atmosphere (Brown et al, 1996; IPCC, 2001a; Freibauer, 2002):

- Conservation to prevent emissions from existing forest carbon pools. This measure has an immediate benefit for the atmosphere. Its theoretical potential equals the current existing carbon stock in forest ecosystems that potentially could be released. Conservation is important in regions with high C stocks per area and where natural disturbances are insufficiently frequent or intensive to cause large immediate reductions in C stocks.
- Sequestration to increase stocks in existing pools. The effect of sequestration can be characterized by a slow build-up of carbon in the litter and soil from tree growth and carbon accumulation. Activities that aim to promote carbon sequestration can lead to carbon gain in the biosphere assuming forest carbon can be restored to the natural carrying capacity. Sequestration applies to areas where C stocks have been depleted.
- Substitution of energy-intensive, or fossil fuel-based products with products derived from renewable resources. The benefits of substitution accumulate over time with each harvest and product use. The technical potential can be as high as the emissions from fossil fuel that potentially can be substituted. The effect of fossil fuel substitution depends on whether the substitution actually reduces fossil fuel use or merely limits its increase. Substitution relies on harvesting, and therefore conflicts with conservation and sequestration objectives in forests.

In plantations, the last two strategies are of particular importance.

Through these measures, forest ecosystem management can introduce and enhance CO_2 sinks to service atmospheric carbon mitigation, affecting the biomass (above and below ground), litter, dead wood, soil organic matter, products and fossil fuel carbon substituted by products, and the use of biomass for energy production (see Figure 3.1). Single management activities can either increase or decrease carbon pools. The response may differ between stand and landscape levels, or be perceived differently by ecosystem and forest sectors. The overall net effect of management is expressed by changes in atmospheric carbon stocks. Yet forests are also vulnerable to climate change, and carbon stocks accumulated over decades bear a certain CO_2 efflux potential through various types of disturbances (Körner, 2003).

The short-term effects of management on the forest carbon budget are relatively easy to measure in the above ground biomass. One major challenge is to quantify the long-term effects of historical forest use on soil carbon stocks, and to separate these from the effects of recent forest management.

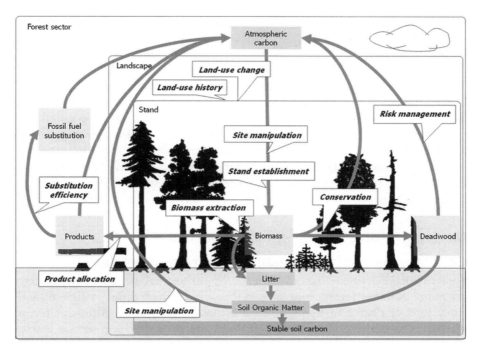

Figure 3.1 *Overview of main effects of forest management options for mitigation*

Note: The chart shows carbon pools and services (grey boxes), and fluxes (arrows). Specific management activities are presented in white boxes (speech bubbles). Management activities can have multiple effects on different pools, all of which cannot be presented here
Source: adapted from Böttcher, 2008

While some practices like inter-rotational site preparation practices can have a very substantial effect on soil C in tropical plantations (Paul et al, 2002), the effects of silvicultural practices on soil carbon are comparably small given the large small-scale variability of soil carbon stocks (Smith and Conen, 2004; Mund and Schulze, 2006).

The carbon balance of the forest-based sector includes both the forest resources and also harvested wood products (e.g. Marland and Schlamadinger, 1997; Liski et al, 2001; Nabuurs et al, 2007; Freibauer et al, submitted). The production of wood is the primary function in 34 per cent of the world's forests. Global wood removals were estimated to be approximately 0.8Gt C in 2005 (FAO, 2005). Wood products affect the carbon cycle in three major ways (IPCC, 2001a):

- as a physical pool of carbon, maintained through recycling, and depleted by decay;

- as an energy or material substitute replacing fossil fuels and energy-intensive materials;
- as an energy source.

This chapter reviews the relationship between management options in planted forests and carbon sequestration under present and future climate conditions. We review the state of knowledge about the effects of new forest plantations, and of changes in management practices in existing plantations on the carbon balance. Potential impacts of climate change on the carbon balance are summarized and consequences for forest management are outlined. Finally, we discuss the potential synergies and trade-offs between carbon management options and the provision of other goods and services from forests.

Managing forests for carbon sequestration

General effects of forest management on the carbon balance

Forest management significantly affects the carbon balance in forests. There are many different forest types around the world, which differ in climate, soil conditions or degree of human influence on the ecosystem. Although forest management varies greatly, certain features are nevertheless common across forest ecosystems.

In most cases, primary unmanaged forests contain significantly higher carbon stocks than managed forests, in particular short-rotation plantations (Cannell, 1995; WBGU, 1998). However, carbon storage in managed forests can exceed that in unmanaged stands if, to avoid disturbance, the harvest cycle is extended beyond the average length of the natural disturbance cycle (Price et al, 1997). Whereas in regions where natural disturbance events occur relatively infrequently, old-growth forests store significantly larger quantities of biomass carbon than younger managed stands, net annual carbon sequestration rates in managed stands are usually higher. In Figure 3.2 patterns of stand-level carbon accumulation over time are compared between an unmanaged forest and forests, including plantations, under different management regimes. Carbon sequestration rates are high in young forests and decline as forests grow older. Assuming an undisturbed development, biomass carbon stocks increase constantly until they reach a maximum value. High stocks are accompanied by small increments. The decline in carbon stocks varies with management intensity. Simultaneously, carbon sequestration rates increase with management intensity. The graphs (a) to (d) in Figure 3.2 also demonstrate the importance of temporal fluctuations in carbon stocks at the stand level.

Converting old-growth forests with high C stocks and low C sequestration rates into young, fast-growing plantations with high carbon sequestration, however, has a negative impact on the net greenhouse gas balance because the large initial loss of carbon cannot be compensated for within a conceivable period of time by the additional carbon sequestration in the growing

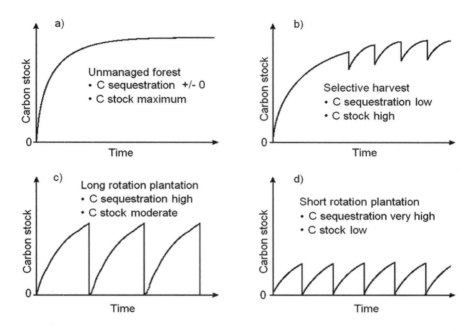

Figure 3.2 *Carbon stock development in forests under different management systems*

Source: after WBGU, 1998

plantation and harvested wood products (Harmon et al, 1990; Kurz et al, 1998; Schulze et al, 2000).

The integrated effect of alternative management systems is best recognized at the landscape level. By extending the view beyond the forest stand to the level of forest management units or forest landscapes, time-dependent carbon gains and losses average out. Then, implications of different silvicultural systems for C storage and sequestration become more apparent.

The initial age structure of the forest landscape is an important key of carbon dynamics in planted forests at the landscape level (Böttcher et al, 2008). The age-class structure itself reflects the past afforestation rate, harvesting activities and stand-replacing disturbances. As discussed above, older forests usually contain more carbon than young stands. Therefore different management regimes, e.g. with different rotation lengths, can affect landscape carbon stocks differently when applied to a forest landscape depending on the distribution of age classes in the plantation landscape. This legacy effect might affect whole regions, e.g. as in Canada. Shifts in the age-class structure of Canadian forests as a result of changes in stand-replacing disturbance regimes caused the Canadian forests to become a net source of carbon in the 1980s after being a strong sink for decades (Kurz and Apps, 1999).

In general, the overall effectiveness of carbon management as a greenhouse gas mitigation activity in the land use sector depends on the initial status of the ecosystem and the proportional cover of the various forest stages in the landscape. In degraded areas with an open canopy, restoration can result in large carbon gains per hectare (Nabuurs et al, 2007). In contrast, especially in regions with ongoing deforestation, the conservation of carbon stock through the protection of primary forests offers mitigation potential. Similarly, the carbon sequestration potential in young forest landscapes with sustainable forest management with long rotations is large, whereas the management of forests with a high proportion of old stands needs to aim at maintaining these stocks. Increasing the forest cover by the afforestation of unproductive land area is one of the most efficient long-term carbon management options.

Creating new forests

The notion of storing CO_2 by planting trees was developed as early as the 1970s (Dyson, 1977) and proposed again when the climate change discussion became more public following the 1992 Earth Summit in Rio de Janeiro (Marland and Marland, 1992). The development of a global afforestation programme to sequester carbon in plantations was proposed (Nilsson and Schopfhauser, 1995). At that time, it was estimated that this could result in 345Mha of new forest plantations and agroforestry plantings, which would sequester up to 1.5Gt C per year, equivalent to about 30 per cent of the anthropogenic carbon emissions. A decade later, although afforestation for CO_2 sequestration is still of interest, its expected contribution of carbon is significantly smaller. The IPCC Third Assessment Report estimated that 12–15 per cent of fossil fuel emissions up to 2050 could be offset by improved management of terrestrial ecosystems globally (Sathaye and Bouille, 2001). A more recent regional study in the Midwestern United States reported that the afforestation of marginal agricultural land, comprising 24 per cent of agricultural land area, could offset 6–8 per cent of current CO_2 emissions from the regional fossil fuel combustion (Niu and Duiker, 2006).

To assess the net carbon mitigation impact of an afforestation programme, consideration must be given to the balance of deforestation and afforestation. For example, Woodbury et al (2006) calculated that, in the period 1990–2004, deforestation in the south-central and south-eastern US resulted in the release of 49Tg C from the soil; more than 50 per cent of the 88Tg C in soils sequestered by afforestation in the region during this time period. The net carbon sink in the tree biomass was about 240Tg C, resulting from a much larger afforested area compared to the deforested area. However, globally, the latest Forest Resource Assessment by the United Nations Food and Agriculture Organization (FAO, 2005) estimates the current rate of afforestation, landscape restoration and natural expansion of forests at 5.7Mha per year, in contrast to a loss of 13Mha per year through deforestation.

The area of forest plantations increased by 2.8Mha per year between 2000 and 2005, mostly in Asia (FAO, 2005). According to the Millennium

Ecosystem Assessment scenarios (MEA, 2005), the forest area in industrialized regions will further increase between 2000 and 2050 by about 60–230Mha. These areas of new forests are characterized by a very juvenile age structure (e.g. as for the old coppices or post-war plantations of European forests) that show high increment rates and increasing growing stock. The sustained accumulation of carbon in these forests also results from harvest rates that are lower than the increment because harvesting has not adapted to the increase in productivity (Ciais et al, 2008).

Afforestation typically leads to increases in biomass and dead organic matter carbon pools and, to a lesser extent, soil carbon pools (Paul et al, 2003). However, biomass removal and site preparation prior to afforestation also may lead to carbon losses. The carbon sequestration potential of afforestation depends on many different factors such as previous land use, soil type or tree species (Guo and Gifford, 2002; Jandl et al, 2007). Carbon sequestration is higher on sites deplete in soil carbon, e.g. due to unsustainable agricultural practices (De Koning et al, 2005; Jandl et al, 2007). Conversely, on sites with high initial soil carbon stocks, (e.g. grassland ecosystems), soil carbon stocks may decline following afforestation (Davis and Condron, 2002; Thuille and Schulze, 2006). However, depending on the ecosystem type and productivity, more often the accumulation of carbon in the forest biomass compensates for this loss after about 10–15 years (Thuille and Schulze, 2006). The degraded soils on land available for afforestation in the tropics primarily constitute low-carbon systems where a significant net C sink after afforestation is expected. Grasslands with high initial carbon stocks that more likely lead to C release after afforestation are more often associated with dry or cold/wet climates.

Thuille and Schulze (2006) described the development typical of different ecosystem C pools following an afforestation of former grasslands (Figure 3.3). After tree planting, carbon is released from the mineral soil for at least several years, whereas carbon is sequestered mainly in the above ground biomass. With increasing tree biomass in regions where decomposition is restricted by dry and/or cool periods, litter production increases, building up an organic layer and, in the longer term, the carbon pool, at least in the top soil, is partly replenished (Zerva and Mencuccini, 2005).

In their comprehensive review of the change in soil carbon after afforestation, Paul et al (2002) found a clear age effect, indicating carbon losses in most young afforestations, variable trends in 5–30-year-old plantations and prevailing soil carbon gains 30 years and more after afforestation. Furthermore they documented species differences: in surface soil (<10cm or <30cm), C amounts increased under hardwoods (poplar, mahogany, etc.) and softwoods (mixed pines, spruce, etc.), yet changed little under eucalypts and decreased under radiata pine. In deeper soil layers (>10cm depth), C increased under eucalypts and other hardwoods, and decreased under radiata pine and other softwoods. It should be noted that the different species were planted in different climatic regions of the world. However,

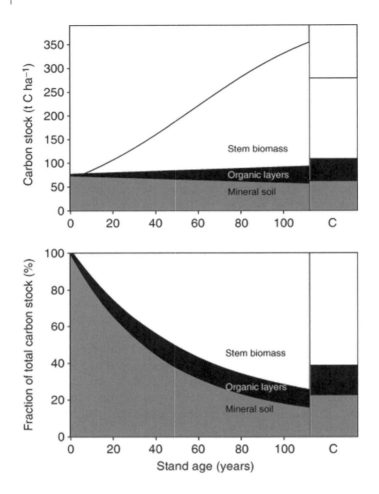

Figure 3.3 *Changes in carbon stocks in stem biomass, organic layers and mineral soil to 50cm depth, averaged over several chronosequences*

Note: Previous land use was pasture. The bars on the right hand side show the carbon stocks in continuously forested control plots
Source: adapted from Thuille and Schulze, 2006

species-specific responses are also documented from targeted experiments by planting several species along a site fertility gradient (e.g. Vesterdal and Raulund-Rasmussen, 1998).

We can conclude that creating new forests results, in most cases, in significant carbon sinks. However, the influence of former land use and selected species stresses the importance of management choices for optimal carbon management of these areas.

Changing management of existing forests

In the following, different management options in existing forests are reviewed with respect to their impact on the carbon budget and potential mitigation contribution. These options include choice of rotation length, thinning intensity, enhancement of tree growth, choice of tree species, treatment of harvest residues, protection from disturbances and management of harvested wood products. Figure 3.4 presents an overview of the estimated sequestration potential of a range of forest management measures. This suggests that temporal changes in forest carbon stocks could range from 0.15t C/ha/year to 3.5t C/ha/year (Nabuurs et al, 2000).

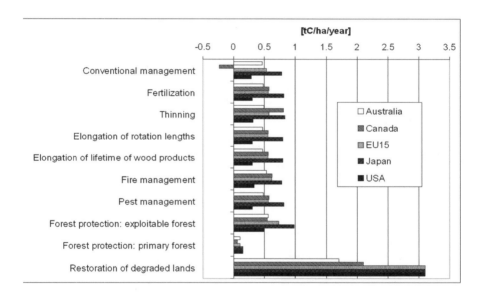

Figure 3.4 *Total potential sequestration per hectare (high estimates) for different forestry activities*

Source: Nabuurs et al, 2000

Rotation length

Rotation length (i.e. the time from stand establishment to harvest) is commonly used to manage timber yield and income from forests and directly influences carbon stocks in biomass, soil and wood products (Liski et al, 2001; Harmon and Marks, 2002; Pussinen et al, 2002; Kaipainen et al, 2004). Changing the rotation length can therefore be considered an effective measure for managing the carbon budget of forests for climate change mitigation. Kaipainen et al (2004) investigated the effect of rotation length in different

European forests on carbon stocks in forest biomass, soil and wood products using the CO2FIX model. Biomass carbon was affected most. An increase in rotation length usually results in an increase in biomass carbon stocks (Figure 3.5).

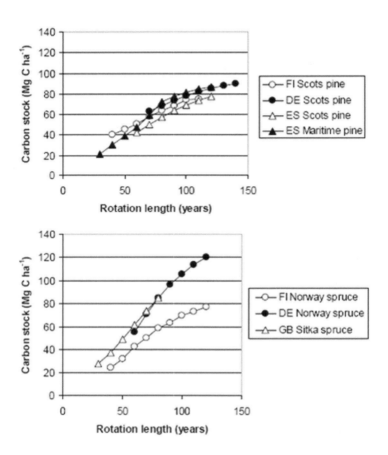

Figure 3.5 *Carbon stock of trees in forests in Finland (FI), Germany (DE), Spain (ES) and the UK (GB) for different rotation lengths*

Source: adapted from Kaipainen et al, 2004

The effects of rotation length on the amount of carbon stored in soil are less well understood (Johnson, 1992; Johnson and Curtis, 2001) and more difficult to assess. Prolonging rotation length may increase soil carbon due to (a) more living biomass and consequently an increase in litter input from tree biomass and (b) less frequent harvests, which may stimulate decomposition through organic matter mixing with the top layer of the mineral soil (Yanai et al, 2003). However, more frequent harvests also increase litter input and the amount of soil carbon (Seely et al, 2002). Similarly, Liski et al (2001) report increases in carbon stored in soil with rotation shortening after modelling both Scots pine and Norway spruce growth over 60 as well as the recommended 90 years. The soil carbon stock increased with the increased litter fall from young trees which, unlike old trees, tend to allocate more growth to tree parts other than stems, and leave more harvest residues per harvested stem wood volume (Liski et al, 2001).

The effect of rotation length on carbon stocks in harvested wood products depends on both the quantity and the quality of harvested timber (Kaipainen et al, 2004). While a longer rotation might reduce the harvested yield per hectare due to losses through natural mortality (Figure 3.6), the average carbon stock of wood products might remain stable or even increase. This is because the mean residence time of carbon in products increases when a larger proportion of the wood products manufactured derive from larger-sized timber with a longer lifespan (Kaipainen et al, 2004).

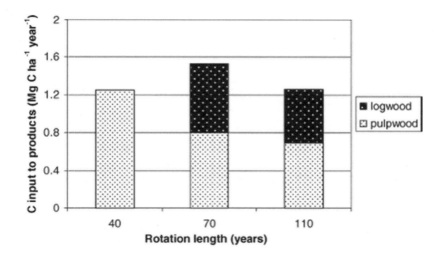

Figure 3.6 *Mean annual harvests and wood assortments from Scots pine forests in Finland for different rotation lengths*

Source: adapted from Kaipainen et al, 2004

Whether rotation prolongation is favourable for climate change mitigation depends on current rotation length, tree species, associated changes in wood products and the occurrence of natural disturbances. The effect on the landscape harvest yield depends on the position of the culmination point of the stand mean annual increment. Any change in rotation length away from the optimal rotation length leads to a decrease in harvest yields. However, few forest landscapes have a balanced age-class structure. Moreover, forest growth has increased substantially in many forests in the Northern hemisphere due to nitrogen deposition, CO_2 fertilization and climate change (Magnani et al, 2007; Kahle et al, 2008). Thus, in many regions, removals have been well below net annual increments in the late 20th century (UNECE/FAO, 2000) and, consequently, the average rotation length increased widely after 1990 with no constraints on harvest level.

The rotation time in managed forests often is influenced by the potential product the forest owner wants to sell and the current market situation for different products. How effectively rotation extension is implemented in the future to mitigate climate change is dependent to a large degree on economic incentives (see Figure 3.7). By increasing the rotation length in Finnish spruce forests by 30 per cent, Liski et al (2001) calculated losses of landowner mean net revenues of US$14.9/ha/year or 10 per cent, as a result of the decreased annual fellings. Over the next 20 years, renewable energy policies are expected to encourage the establishment of short-rotation forest plantations for wood fuel production (UNECE, 2005). In addition there is a trend towards compound products, resulting in a higher demand for sawn timber of smaller diameters (UNECE, 2007). It is therefore likely that economic conditions will favour a reduction in rotation length in future in some forest management regions.

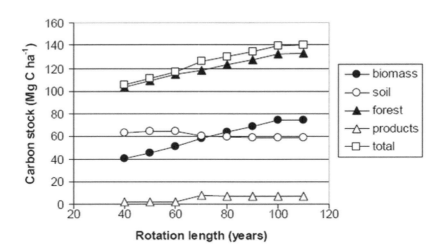

Figure 3.7 *Mean carbon stocks over the rotation of Scots pine forests in Finland for different rotation lengths*

Note: Biomass includes the living biomass of trees (excl. fine roots), soil includes litter and soil organic matter, forest includes biomass and soil, products means wood products in use and total is the sum of biomass, soil and products
Source: adapted from Kaipainen et al, 2004

The effect of extending rotation length is only temporary. Once forests reach a steady state with this increased rotation length, they are no longer carbon sinks, and the lower the revenue or the higher the roundwood prices are, the higher the costs of maintaining the larger carbon stock (Liski et al, 2001). At the same time the higher carbon stocks also imply an increased risk of unintended emissions from disturbances such as storms or fire.

Thinning intensity

In many forestry regions, forest stands are thinned, providing early revenues and making site resources available to fewer individual trees. Optimal timing and intensity of thinning increases the volume of individual trees compared to unthinned stands (Nyland, 1996). However, at the stand level, thinning reduces tree density and therefore reduces carbon in the biomass pool. Between well-timed thinning operations, the number of trees remains unchanged, meaning that thinning mimics self-thinning processes. This reduces the loss of carbon through natural decomposition and leaves more carbon for harvest, wood products and substitution. Depending on tree species, the trees remaining after a heavy thinning that reduces stand density below a critical level cannot compensate for the loss in crown cover of the removed trees at the stand level (Assmann, 1968), and biomass increment also declines.

Modelling exercises indicate that any thinning regime decreases forest ecosystem carbon stocks compared to unthinned stands (e.g. Dewar and Cannell, 1992; Thornley and Cannell, 2000; Eriksson, 2006). In addition to reducing average carbon storage in biomass pools, intensive thinning can impact negatively on carbon stored in litter and dead wood pools (Seely et al, 2002). Therefore, forest ecosystem carbon storage could be maximized by abandoning thinning, yet would be associated with a substantial loss of revenue for forest owners (Garcia-Gonzalo et al, 2007).

Böttcher et al (2008) show how increasing thinning intensity (with more biomass ending in slash or relatively short-lived wood products) might reduce the carbon balance, even at the forestry sector level when harvested wood products are considered. The fate of the harvested wood is important in the assessment of the net effect of intensifying thinning. Even if the mean residence time of carbon stored in wood products from early thinning operations were shorter than the mean residence time of carbon in forest biomass or dead wood – which is not true for modern compound wood products – the efficient use of thinned wood nevertheless would contribute to climate change mitigation through the product recycling chain and substitution effects (Marland and Schlamadinger, 1997; Freibauer et al, submitted).

Thinning is an important measure for managing species distribution in mixed stands and thus can indirectly affect carbon stocks through species choice. It may also reduce the risk of disturbances, such as storms, fire or snow breakage in younger stands, and therefore could be considered a disturbance protection measure (see below) that leads to decreased stocks in the short term but assures stable stands and carbon stocks in the long term.

Enhancing biomass growth

Forest stand growth is determined by (a) site quality (the main factors are climate, microclimate, soil and relief) and (b) how well trees capture resources. The latter is influenced by genetic pool, stand development and degree of stocking. The genetic stock and other silvicultural factors, most of which can be altered by managers, may prevent the potential site productivity from being attained (Mead, 2005). On sites where nutrients and water are limiting factors to plant growth, the potential for enhancing and accelerating biomass production and yield by fertilizer application and irrigation may be considerable (e.g. Nilsson, 1997; Stromgren and Linder, 2002; Adams et al, 2005; Bergh et al, 2005; Mead, 2005). However, whether fertilization also results in enhanced long-term C sequestration (Pettersson and Hogbom, 2004) is less clear and probably only valid in some cases.

Figure 3.8 lists net responses in productivity of forest plantations to various growth improvement measures reviewed by Mead (2005). The review considers optimal species and provenance selection assumed to be suitable for the site. Responses to treatments are always very site-dependent. Furthermore, some treatments may not be applicable to particular sites, nor are effects always cumulative (Mead, 2005). Whereas N and P application increased tree growth in various eucalypt species trials in India (Sankaran et al, 2004) by up to 70 per cent and 43 per cent, respectively, and early weed control increased growth by up to 138 per cent, the establishment of N-fixing legumes had no significant effect on growth. However, in general, depending on growth-limiting factors other than N availability such as N-fixing, unlike in monocultures, a species in mixed-species plantations has the potential to increase productivity while maintaining soil fertility (Forrester et al, 2006, 2007, see also the next section on species choice).

Site preparation for enhancing biomass growth through manual, mechanical and chemical measures and prescribed burning promotes rapid establishment, early growth and the good survival of seedlings but may lead to soil carbon losses, which typically increase with the intensity of the soil disturbance (Jandl et al, 2007). For a thorough assessment of enhanced growth as a measure of climate change mitigation, the effects (of enhanced growth) on soil carbon as well as the greenhouse gas emissions associated with site preparation and fertilizer production need to be taken into account.

Choice of tree species

The choice of tree species for a certain site is a management decision and can influence C storage over time. In general, the carbon storage potential of a given species depends on the maximum achievable stand biomass and on the time required to reach that maximum. Fast-growing species usually reach their maximum stand biomass relatively quickly, whereas, although slow-growing species may require more time, they have a higher maximum. These different properties of forest tree species theoretically can be exploited to direct mitigation strength and timing in different ways depending on the anticipated

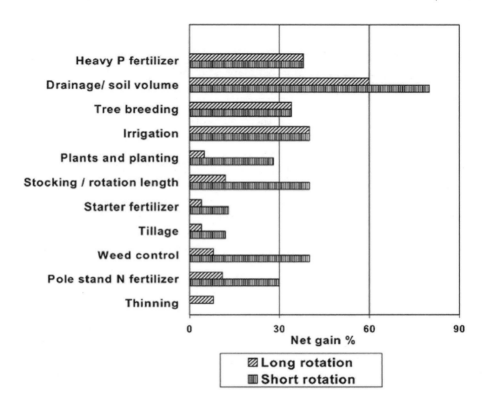

Figure 3.8 *Net responses in productivity to various silvicultural treatments in short-rotation (<12–15 years) and longer-rotation tree plantations*

Source: adapted from Mead, 2005

mitigation strategy. If the objective is to avoid the implications of the cumulative effect of increased CO_2 and temperature, planting slow-growing trees yielding long-lasting wood products would help to reduce climate change impacts. Changing to fast-growing plantations with high product yields and with high substitution efficiencies can be a tool to store carbon rapidly in the short term (Dewar and Cannell, 1992).

Species mixtures also affect productivity substantially. Forrester et al (2004) compared mixed and mono-specific stands of *Eucalyptus globulus* and *Acacia mearnsii*. At the tree level, eucalypt and acacia heights, diameters, volumes and above ground biomass were higher in mixtures than in monocultures a few years after planting, indicating that biomass accumulation in *E. globulus* plantations can be increased by acacia admixture. However, competition for resources other than nitrogen, such as light, soil moisture or other nutrients may balance positive mixing effects, e.g. through increased N availability (Forrester et al, 2007).

In addition, tree species change can have effects on soil carbon through litter quality (Giardina et al, 2001; Binkley and Menyailo, 2005), nitrogen fixation (Resh et al, 2002) and also rooting patterns (Johnson, 1992). By influencing water balance, microclimate and nutrient availability in soils (Augusto et al, 2002), choice of tree species also indirectly affects soil carbon stocks.

Tree species also differ in their susceptibility to biotic and abiotic disturbances. Converting pure stands of coniferous species into mixed stands with broadleaved trees may therefore reduce carbon losses due to higher stand stability (Spiecker, 1999; Jandl et al, 2007). Based on a quantitative review of over 50 comparative studies on insect pests, Jactel et al (2005) demonstrated that tree species growing in mixed stands overall suffer less pest damage, or have lower pest populations than pure stands. However, one exception was evident, whereby polyphagous pest insects build up their populations on a preferred host tree species initially and then spill over on to associated host tree species.

Treatment of harvest residues

Besides rotation length and amount of biomass harvested, the treatment of detritus or harvest residues impacts on forest carbon stocks in managed forests (Harmon and Marks, 2002; Laiho et al, 2003; Eriksson et al, 2007). The treatment may comprise various alternatives such as slash-burning, removal for bioenergy use or decomposition on site. Decomposition leads to a gradual carbon loss from the ecosystem, sometimes even offsetting the biomass carbon accumulation rate leading to a negative net balance. This is especially true if the material inhibits tree regeneration. Alternative treatments like slash-burning lead to a much faster loss that is probably similar in magnitude over time. Biomass removal for use as fuelwood or bioenergy causes rapid losses to the ecosystem but affords opportunities for fossil fuel substitution outside the forest (Palosuo et al, 2008).

Laiho et al (2003), in keeping with Johnson et al (2002) and Sanchez et al (2006), found that differences in litter carbon triggered by different logging residue treatments do contribute mainly to short-term differences in soil carbon. Tiarks and Ranger (2006) also observed, in a review of 16 sites in tropical and subtropical regions, that the effects of clear-cutting and replanting and the effects of slash management on soil C are minor. These findings contrast with those from other authors who found slash treatments significantly affected soil C after looking at variables that might indirectly influence carbon storage in soils in the short and long term. On sites where the slash is left on the forest floor, soil moisture was found to increase (O'Connell et al, 2004). After whole-tree harvesting and the removal of harvest residues from the ecosystem, the supply of nutrients steadily declines (Rosenberg and Jacobson, 2004; Merino et al, 2005; Gonçalves et al, 2007). The latter effect clearly has long-term implications for tree growth and biomass carbon storage. At sites where nutrient stores are naturally low, nutrient export through wood harvest can be large (Deleporte et al, 2008). In the Congo, for example, they found the total tree harvest removed about 22 per cent of soil Ca, retaining slash and debarking on site removed only 8 per cent of available Ca in the soil.

The treatment of slash material may also have implications for an ecosystem's disturbance regime, e.g. by altering susceptibility to fire, a mitigation option that is discussed in the section below.

Protection from disturbances

Although old forests can achieve significant carbon accumulation rates (Schulze et al, 2000; Knohl et al, 2003), forest growth declines with stand age (Ryan et al, 2004). Due to stand-replacing disturbances, many forests even become sources of CO_2 after maturity (Körner, 2003). Given the life expectancy of trees (commonly 50–300 years) and the non-random mix of age classes, the proportion of growing forests slowly accumulating carbon in most forest landscapes is very high (98.0–99.7 per cent of forest land) and those emitting carbon very low (0.3–2 per cent; Körner, 2003). In these areas, forest management priorities for climate change mitigation range from sequestration to conservation through to protection from disturbances.

One option for wildfire hazard abatement is to use prescribed burning to reduce fuel loads (e.g. Fernandes and Botelho, 2003; Narayan et al, 2007). Other options for reducing fuel load in plantations include proper tree maintenance and tree spacing (Saharjo, 1997) as well as the establishment of fire breaks, that is ploughed strips where vegetation is removed completely (Franklin, 2001). Fuel management typically reduces biomass in the forest at stand level. This in return reduces fire probability and intensity, and reduces total emissions from forest fires. Thus, at landscape level, carbon stocks are usually increased through fuel management due to successful fire suppression (Houghton et al, 2000). Forest fire suppression without fuel management may increase carbon stocks in the short term. However, as recently demonstrated in North America, these measures do increase the risk of devastating large fire disturbances with massive carbon release (Volney and Fleming, 2000).

Understanding the interrelationship between landscape forest patterns and disturbance risks can be important in forest planning for reducing forest fire risk (Amiro et al, 2001; Hirsch et al, 2001; González, 2006). Similar principles have been used for a long time in traditional forest management planning to reduce wind damage in managed forest landscapes (Gardiner and Quine, 2000).

Fate of harvested wood products

Thinning or harvest operations partially replace natural mortality in a forest ecosystem and result in wood products that contain a given portion of the carbon originally stored in the biomass. The consideration of wood products as a carbon pool can resolve the conflict between wood production and carbon sequestration, which has been often reported (e.g. Fischlin, 1994). Several models are available that include carbon storage in forests and wood products, and which have been widely used to calculate C flows in plantations (e.g. Karjalainen, 1996; Masera et al, 2003). The carbon stored in harvested wood destined for use as sawn-wood and wood-based panels, which to a large extent are used in permanent constructions, is bound in these materials over decades. Other products, like fuelwood and paper, store carbon for a few years at the most. The IPCC Third Assessment Report on Mitigation (IPCC, 2001a) lists four options that influence the effect of carbon stored in wood products:

- the level of production of wood products, which may increase or decrease the pool size;
- a change in the quality of wood products, which lead to changes in the average lifetime of products;
- changes in processing efficiency;
- changes in recycling rates and the fate of wood products.

In view of the strong policy incentives, the use of bioenergy from forest biomass to substitute for energy from fossil fuels will increase further in the future (UNECE, 2007). Net effects depend on substitution efficiency, i.e. on how bioenergy is produced, on the type of energy that is substituted and on the extent of displacement (Schlamadinger and Marland, 1994). The production of renewable raw materials in managed forests with subsequent reuse, and the use of residues for energy, is beneficial for climate protection and also economically attractive (Freibauer et al, submitted). Reusing wood products for energy is also an argument against the switch from timber production to energy forestry because, with energy substitution, carbon can be stored in harvested wood products longer. The fate of harvested wood products is a particularly important option for plantations, since, although they comprise less than 5 per cent of the global forest area, they produce more than 40 per cent of the wood harvested worldwide (FAO, 2005).

Carbon and forest management under climate change

Expected climate change and impact on forests

Climate is an important driver of forest development and ecosystem processes. Consequently, climate change will have multiple direct and indirect effects on the forest carbon balance. Climatic conditions have already changed significantly: global average temperature has increased by 0.8°C since 1900 (Hansen et al, 2006) and the 12 hottest years observed globally since 1880 all occurred between 1990 and 2005. The recent European heat wave of 2003 was a drastic demonstration of the extent of impacts we can expect more often in the future (Schär and Jendritzky, 2004; Ciais et al, 2005). Different climate-related factors are expected to change: the atmospheric CO_2 concentration will increase further to at least twice the pre-industrial concentration of 280ppm, temperatures will rise unevenly in different world regions with likely changes in seasonal and daily amplitudes, precipitation amount and distribution will affect water regimes, and various disturbance regimes (storm, fire, snow, insects and diseases) will be either directly or indirectly affected as well.

The expected impacts of the climatic factors on forests include changes in species-specific growth rates, productivity and mortality, all of which will affect the competitiveness of individual species and species distributions. Regional climate change impacts on forests will reveal large differences ranging from growth enhancement in Northern Europe and other boreal regions to mainly negative impacts, for example in the Mediterranean region.

Concerns have been raised about the permanence of the current terrestrial carbon sink in the northern hemisphere (Ciais et al, 1995). A main source of uncertainty is the response of vegetation and soil carbon to global change (Friedlingstein et al, 2003). Two critical questions are how the decomposition of soil organic matter will respond to increasing temperatures and whether the huge soil carbon pools, especially in high latitudes and peat lands, will decline (Giardina and Ryan, 2000; Davidson and Janssens, 2006; Kirschbaum, 2006).

Climate change will also increase climate variability and most probably lead to more frequent and severe extreme weather conditions (IPCC, 2001b, 2007). This will affect both tree growth and mortality directly, which ultimately will affect the natural disturbance regimes as well.

Consequences for carbon management

Adaptive forest management can utilize species-specific characteristics to reduce the risks associated with the adverse effects of climate change and to mobilize potential benefits such as increased production rates. Tree species differ, for example, in growth rates, drought sensitivity and susceptibility to disturbances. Forest management involves making choices about species selection and mixtures, the use of natural or artificial regeneration, the timing and intensity of thinnings and harvesting, the initiation of prescribed burning, the application of fertilizers and even irrigation in some intensive plantation systems. Yet these management decisions again feed back to the carbon budget. A general overview of links between mitigation and adaptation options is given in Table 3.1.

Pest and disease control is common practice in managed forests. Economically important species like Norway spruce in Central Europe have been grown for centuries outside their natural distribution range. Norway spruce forests are found in strictly protected forest areas (e.g. Bayerischer Wald National Park in Bavaria, Germany) where, without the application of pest control measures, it would be very susceptible to bark beetle damage leading to large-scale forest dieback. Mountain pine beetle infestation in Canada is a more recent drastic example of how climate change may severely affect disturbance regimes and ranges (e.g. at higher elevations; Williams and Liebhold, 2002). Management interventions may reduce or delay carbon release through such biotic disturbances, but, in most cases, may be unable to completely eliminate them.

Table 3.1 *Mitigation and adaptation matrix*

Mitigation option	Vulnerability of the mitigation option to climate change	Adaptation options	Implications of adaptation for greenhouse gas emissions
Afforestation/ reforestation	Vulnerable to changes in rainfall, higher temperatures (increase in forest fires, pests, dieback due to drought)	Species mix at different scales. Fire and pest management. Increase species diversity in plantations. Introduction of irrigation and fertilization. Soil conservation.	No or marginal implications for greenhouse gas emissions; positive if the effect of perturbations induced by climate change can be reduced. May lead to increase in emissions from soils or use of machinery and fertilizer.
Forest management in plantations	Vulnerable to changes in rainfall, higher temperatures (increase in forest fires, pests, dieback due to drought)	Pest and fire management Adjust rotation periods. Species mix at different scales.	Marginal implications on greenhouse gases. May lead to increase in emissions from soils or use of machinery or fertilizer.

Source: Nabuurs et al, 2007

As most tree species are long-living, one of the main challenges of forest management in confronting climate change is to decide when to change species preferences. As climate conditions are changing gradually, the competitive relationship between species has also shifted. Yet, in many cases, drastic change occurs as a result of extreme events (cf. drought in 2003; Ciais et al, 2005) not from gradual increases in temperature. Management needs to decide whether to put up with anticipated growth declines of current species in the future, or favour new species expected to be more suited to future conditions. Choosing new species will always be accompanied by risks because suitable genotypes are identified by testing, and only this can determine whether local conditions are suitable for sustaining the growth and reproduction of exotic species. However, suitable species may already be present in the target region, allowing for a much more rapid adaptation of management practices.

Agroforestry management systems also represent an opportunity for synergies between carbon sequestration and adaptation to a changing climate (Verchot et al, 2007). In addition, agroforestry can help to decrease pressure on natural forests and promote soil conservation (Nabuurs et al, 2007).

An important challenge for carbon management under changing climate conditions is the need to cope with uncertainty (e.g. of future regional climate development) and risks (e.g. storm or fire disturbances). In this context, management strategies at the district or landscape level are important. This allows for an increased diversity of species, and the application of alternative management types and response strategies in different stands. Moreover, techniques that alter landscape patterns for fire management purposes (also referred to as fire-smart management) can also reduce the impact of increased disturbance risks (Hirsch et al, 2001). In general, there is a strong need by local decision-makers as well as by forest researchers for detailed reliable regional climate change scenarios and associated uncertainties.

Carbon vs. other goods and services

Important forest goods and services include timber production, biomass for bioenergy, non-timber forest products, carbon storage, water recharge and water quality, nature conservation and the protection of biodiversity, soil protection, amenity values and recreational use. There are obvious trade-offs between certain ecosystem goods and services. Interestingly, it is not possible to maximize different carbon management objectives in the one stand. Growth rates and, consequently, carbon sequestration rates are higher with short-rotation management because stand growth rates decline in mature forest stands. In contrast, carbon storage in the forest is highest in unmanaged forests.

The design of carbon management strategies should consider the trade-offs between these alternative options. Increasing forest ecosystem carbon stocks must be evaluated against increasing the sustainable rate of harvest and transfer of carbon to meet human needs (Nabuurs et al, 2007). The selection of mitigation strategies should minimize net greenhouse gas emissions throughout all sectors affected by these activities. For example, stopping all forest harvesting would increase forest carbon stocks, but would reduce the amount of timber and fibre available to meet societal needs. Other energy-intensive materials, such as concrete, aluminium, steel and plastics, would be required to replace wood products, resulting in higher greenhouse gas emissions (Gustavsson et al, 2006).

Climate-effective measures therefore need to increase incentives for management systems that maintain high average C stocks in the long term, such as long-rotation forestry and conservation, which otherwise are at risk of being lost given short-term economic considerations and other land pressures. These measures do not necessarily need to focus on forests. An effective recycling of wood products, for example, can free land for long-term sustained C sequestration and additional conservation services (Böttcher, 2008).

When evaluating optional forest management strategies in relation to their performance in delivering selected goods and services from the list cited above, it can be seen that some of these are in line with either the objective to increase carbon storage or carbon sequestration, while others are not correlated at all (Figure 3.9). The biodiversity value of forests is often inversely related to management intensity. Unmanaged forests are usually structurally rich and provide habitat for many different species. In contrast, short-rotation monocultures with the highest carbon sequestration rates are structurally much simpler and have lower value for protecting indigenous biodiversity (see Chapter 5).

The protective function of forests, for example the protection of mountainous regions against soil erosion, avalanches or damage to infrastructure, is best achieved by continuous forest cover, often with high carbon storage. As already discussed, timber production is more compatible with carbon sequestration objectives, yet it is possible to achieve considerable carbon storage in management systems with long rotation periods and selective cutting directed typically towards the production of high-quality timber.

Groundwater recharge, another important ecosystem service, is neither supported by high carbon storage nor carbon sequestration because this is achieved best in open forests with low stocking density and low growth rates. Amenity values and recreational forest use are missing from Figure 3.9 as they depend more on infrastructure and the population density in the vicinity of the forest.

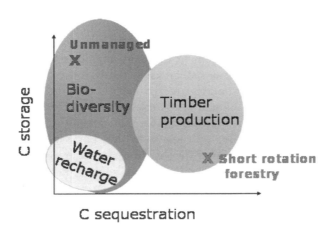

Figure 3.9 *Trade-off between carbon sequestration/storage and other forest ecosystem goods and services. Long rotation length favours carbon storage whereas a short rotation enhances carbon sequestration rates*

Note: Timber production often is associated with moderate to high carbon sequestration rates and reduced carbon storage compared to unmanaged systems. Biodiversity can be high with variable carbon storage, but often declines in intensive, fast-growing management systems. Water recharge is highest with both low carbon storage and sequestration rates in open forest areas.

Another important aspect, related to the possible trade-offs between alternative management options, is the fact that different stakeholder groups have different views regarding the importance of the range of forest ecosystem goods and services. Fürstenau et al (2007) showed how the priorities for different partial management objectives vary among stakeholder groups, and how this affects the utility of alternative management options as perceived by the different groups.

To guarantee the supply of goods and services from forests, the segregation of intensively used productive land and unproductive land for the purpose of C storage and biodiversity was proposed (Huston and Marland, 2003; West and Marland, 2003). Figure 3.10 presents a gradual system of change along a matrix that prioritizes climate-friendly land-use options based on the criteria C stocks currently existing on land, productivity and accessibility. When quantitatively scaled to the region of interest (local to global) it could serve as a tool for land-use planning. Climate-friendly land use would maintain the existing carbon stocks and productivity of the land.

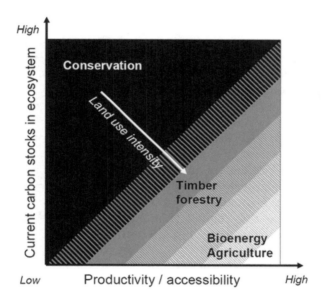

Figure 3.10 *Matrix guiding land management decisions for effective climate change mitigation*

Note: Black = conservation, grey = forestry, white = agriculture
Source: Böttcher, 2008

Plantation carbon management and climate policy

Under the Kyoto Protocol, industrialized countries are committed to reducing their 1990 greenhouse gas emissions from 2008 to 2012 by roughly 5 per cent (UNFCCC, 1997) through their activities in land use, land-use change and forestry (LULUCF), which they nominate. The land-use change from other land uses to new forests (afforestation) or from forests to other land uses (deforestation), the subject of Article 3.3 of the Kyoto Protocol, and their carbon emissions and removals must be reported and accounted for in Annex 1 countries. The role of existing managed forests in the terrestrial carbon cycle was acknowledged by the Conference of the Parties (COP) in the Kyoto Protocol by introducing Article 3, paragraph 4. This article allows countries to choose any forest management, cropland management, grazing land management and revegetation activities to meet their emission reduction commitment. This option was voluntary for the first commitment period of the Kyoto Protocol (2008–2012) and chosen by 20 countries of 39 listed in the Kyoto Protocol Annex 1. However, in the 'Marrakesh Accords', the COP agreed that only fluxes directly induced by human activity would be accounted for, and natural and indirect human-induced effects excluded. The maximum contribution of the forest management sink to the commitment was therefore limited by country-specific caps that were negotiated (Höhne, 2006). The challenge for the integration of forestry activities with climate policy is to design a future accounting scheme for forest management that increases incentives for change towards climate-effective management but, at the same, time reduces accountable removals to those triggered by recent, not past practices or disturbances (Böttcher et al, 2008).

Carbon offset markets that allow trading of certificates for emission reductions or removals exist under both the Kyoto Protocol compliance scheme and as voluntary programmes. Forestry activities in non-Annex 1 countries are addressed in the Kyoto Protocol Article 12. This article describes the Clean Development Mechanism (CDM), which permits project-based activities by Annex 1 countries, which reduce emissions or create additional removals regarded as emission reduction measures, in non-Annex 1 countries. In forestry, eligible activities are restricted to afforestation, reforestation and agroforestry projects.

However, CDM projects in the LULUCF sector face several challenges that, so far, have inhibited widespread implementation. These include issues of permanence, additionality and leakage. The non-permanence of carbon stored by forestry projects, due to the risk of disturbances and later management change, is addressed by temporary crediting and verification every five years. The criteria of additionality needs to be fulfilled for CDMs in general, and excludes plantations established for wood supply or purposes other than carbon mitigation. Furthermore, leakage (e.g. deforestation stopped within but increased outside project boundaries) needs to be avoided. These strict rules for CDM, which led to the introduction of a cap for CDM credits from LULUCF to avoid a swamping of the market (1 per cent of Annex I 1990 emissions), are far from being achieved. The exclusion of LULUCF CDM from large emissions trading schemes (ETS), like the EU ETS, also added another barrier to the implementation of projects.

Forestry projects in the voluntary carbon market have been more successful. Voluntary offset markets function outside the compliance markets, and enable companies and individuals to purchase carbon offsets on a voluntary basis. Many new labels, standards and greenhouse gas registries in addition to CDM were launched recently to guarantee the climate-effectiveness of forestry projects.

A new framework for climate policy that is currently under negotiation will most likely consider the sinks and sources within LULUCF as well, and hence influence the management and establishment of plantations in the future.

Conclusions

Plantations provide a multitude of valuable goods and services. Carbon sequestration is one of these services. Its importance has increased markedly because forests are seen as a key pillar of greenhouse gas mitigation to combat climate change. Five strategies for the contribution of the forestry sector have been identified: (1) increase forest area; (2) increase carbon stock in existing forests; (3) protect existing stocks; (4) increase carbon stored in products (yielding also indirect greenhouse gas mitigation through material substitution); and (5) substitute fossil fuels with bioenergy derived from forest biomass and wood.

While there is great theoretical potential for plantation forestry to contribute in options (1) and (2) through afforestation, extending rotations and refilling depleted soil carbon reservoirs, past degradation and continued human need for resources from land make it difficult to achieve this potential in full or in part. Achieving multiple, often incompatible objectives with land and forest management constitutes a continuous challenge for land-use policy-makers and forest managers, particularly because it requires a close interaction with local communities and various community stakeholder groups. The segregation of different forest functions is likely to gain importance and may be a possible solution to ensure that all goods and services can be maintained at the landscape level. Furthermore, increasing the use of long-lasting wood products and improving recycling rates can reduce the pressure on land resources and permit management for alternative non-carbon services.

In this chapter, we reviewed a broad range of forest management measures and strategies that can help strengthen carbon sequestration in the forest-based sector to mitigate climate change, while still maintaining other ecosystem services like supplying timber and non-timber products, drinking water and space for recreation or wildlife.

Carbon sequestration in the forest-based sector is, to a large degree, a non-permanent strategy. Tree plantations that are established, harvested yet not re-established do not contribute further to carbon sequestration. The sequestration phase is finite, lasting for some decades and gained carbon stocks need to be protected thereafter to keep carbon withdrawn from the atmosphere. Sequestration therefore always needs to be guarded by conservation measures to make mitigation strategies effective. Implications for the overall carbon budget at the stand and landscape levels need to be considered.

Climate change itself, with as yet not fully predictable impacts on forest goods and services – including carbon sequestration – is another challenge, requiring adaptive, non-static management strategies for our future forests. As the anticipated impact of climate change will differ strongly between bioclimatic zones, site conditions and forest types, future management strategies need to be region- and site-specific.

References

Adams, A. B., Harrison, R. B., Sletten, R. S., Strahm, B. D., Turnblom, E. C. and Jensen, C. M. (2005) 'Nitrogen-fertilization impacts on carbon sequestration and flux in managed coastal Douglas-fir stands of the Pacific Northwest', *Forest Ecology and Management*, no 220, pp313–325

Amiro, B. D., Todd, J. B., Wotton, B. M., Logan, K. A., Flannigan, M. D., Stocks, B. J., Mason, J. A., Martell, D. L. and Hirsch, K. G. (2001) 'Direct carbon emissions from Canadian forest fires, 1959–1999', *Canadian Journal of Forest Research-Revue Canadienne De Recherche Forestiere*, no 31, pp512–525

Assmann, E. (1968) 'Zur Theorie der Grundflächenhaltung', *Forstwissenschaftliches Centralblatt*, no 87, pp321–220

Augusto, L., Ranger, J., Binkley, D. and Rothe, A. (2002) 'Impact of several common tree species of European temperate forests on soil fertility', *Annals of Forest Science*, no 59, pp233–253

Bergh, J., Linder, S. and Bergstrom J. (2005) 'Potential production of Norway spruce in Sweden', *Forest Ecology and Management*, no 204, pp1–10

Binkley, D. and Menyailo, O. (eds) (2005) 'Tree species effects on soils: Implications for global change', *NATO Science Series 55*, Kluwer Academic Publishers, Dordrecht

Bolin, B. and Sukumar, R. (2000) 'Global perspective', in R. T. Watson, I. R. Noble, B. Bolin, N. H. Ravindranath, D. J. Verardo and D. J. Dokken (eds) *Special Report on Land Use, Land-Use Change and Forestry*, Cambridge University Press, Cambridge

Böttcher, H. (2008) 'Forest management for climate change mitigation: Modeling of forestry options, their impact on the regional carbon balance and implications for a future climate protocol', Fakultät für Forst- und Umweltwissenschaften, Albert-Ludwigs-Universität, Freiburg

Böttcher, H., Kurz, W. A. and Freibauer, A. (2008) 'Accounting of forest carbon sinks and sources under a future climate protocol: Factoring out past disturbance and management effects on age-class structure', *Environmental Science & Policy*, no 11, pp669–686

Brown, S., Shvidenko, A., Galinski, W. R., Houghton, A., Kasischke, E. S., Kauppi, P., Kerz, W. A., Nalder, I. A. and Rojkov, V. A. (1996) 'Working group summary: Forests and the global carbon cycle – past, present and future role', in *Forest Ecosystems, Forest Management and the Global Carbon Cycle*, Springer-Verlag, Heidelberg, pp199–208

Canadell, J. G., Kirschbaum, M. U. F., Kurz, W. A., Sanz, M.-J., Schlamadinger, B. and Yamagata, Y. (2007a) 'Factoring out natural and indirect human effects on terrestrial carbon sources and sinks', *Environmental Science Policy*, doi:10.1016/j.envsci.2007.01.009

Canadell, J. G., Pataki, D. E., Gifford, R., Houghton, R. A., Luo, Y., Raupach, M. R., Smith, P. and Steffen, W. (2007b) 'Saturation of the terrestrial carbon sink', in J. Canadell, D. Pataki and L. Pitelka, *Terrestrial Ecosystems in a Changing World*, The IGBP Series, Springer-Verlag, Berlin, Heidelberg.

Cannell, M. (1995) 'Forests and the global carbon cycle in the past, present and future', European Forest Institute, Joensuu

Ciais, P., Tans, P. P., Trolier, M., White, J. W. C. and Francey, R. J. (1995) 'A large northern-hemisphere terrestrial CO_2 sink indicated by the C-13/C-12 ratio of atmospheric CO_2', *Science*, no 269, pp1098–1102

Ciais, P., Reichstein, M., Viovy, N., Granier, A., Ogee, J., Allard, V., Aubinet, M., Buchmann, N., Bernhofer, C., Carrara, A., Chevallier, F., De Noblet, N., Friend, A. D., Friedlingstein, P., Grunwald, T., Heinesch, B., Keronen, P., Knohl, A., Krinner, G., Loustau, D., Manca, G., Matteucci, G., Miglietta, F., Ourcival, J. M., Papale, D., Pilegaard, K., Rambal, S., Seufert, G., Soussana, J. F., Sanz, M. J., Schulze, E. D., Vesala, T. and Valentini R. (2005) 'Europe-wide reduction in primary productivity caused by the heat and drought in 2003', *Nature*, no 437, pp529–533

Ciais, P., Schelhaas, M. J., Zaehle, S., Piao, S. L., Cescatti, A., Liski, J., Luyssaert, S., Le-Maire, G., Schulze, E. D., Bouriaud, O., Freibauer, A., Valentini, R. and Nabuurs, G. J. (2008) 'Carbon accumulation in European forests', *Nature Geoscience*, no 1, pp425–429

Davidson, E. A. and Janssens, I. A. (2006) 'Temperature sensitivity of soil carbon decomposition and feedbacks to climate change', *Nature*, no 440, pp165–173

Davis, M. R. and Condron, L. M. (2002) 'Impact of grassland afforestation on soil carbon in New Zealand: A review of paired-site studies', *Australian Journal of Soil Research*, no 40, pp675–690

De Koning, F., Olschewski, R., Veldkamp, E., Benítez, P., López-Ulloa, M., Schlichter, T. and De Urquiza, M. (2005) 'The ecological and economic potential of carbon sequestration in forests: Examples from South America', *Ambio*, no 34, pp224–229

Deleporte, P., Laclau J. P., Nzila, J. D., Kazotti, J. G., Marien, J. N., Bouillet, J. P., Szwarc, M., D'Annunzio, R. and Ranger, J. (2008) 'Effects of slash and litter management practices on soil chemical properties and growth of second rotation eucalypts in the Congo', in E. K. S. Nambiar (ed.) *Site Management and Productivity in Tropical Plantation Forests: Workshop Proceedings*, 6–9 November 2004, Piracicaba, Brazil and 6–9 November 2006, Bogor, Indonesia, CIFOR, Bogor, Indonesia, pp22–26

Dewar, R. C. and Cannell, M. G. R. (1992) 'Carbon sequestration in the trees, products and soils of forest plantations: An analysis using UK examples', *Tree Physiology*, no 11, pp49–71

Dyson, F. J. (1977) 'Can we control the carbon dioxide in the atmosphere', *Energy* (UK), no 2, pp287–291

Eriksson, E. (2006) 'Thinning operations and their impact on biomass production in stands of Norway spruce and Scots pine', *Biomass and Bioenergy*, no 30, pp848–854

Eriksson, E., Gillespie, A. R., Gustavsson, L., Langvall, O., Olsson, M., Sathre, R. and Stendahl, J. (2007) 'Integrated carbon analysis of forest management practices and wood substitution', *Canadian Journal of Forest Research*, no 37, pp671–681

FAO (2005) *Global Forest Resources Assessment 2005*, FAO Forestry Paper 147, Food and Agriculture Organization of the United Nations, Rome

Fernandes, P. M. and Botelho, H. S. (2003) 'A review of prescribed burning effectiveness in fire hazard reduction', *International Journal of Wildland Fire*, no 12, pp117–128

Fischlin, A. (1994) 'Conflicting objectives while maximising carbon sequestration by forests', in M. Apps and D. T. Price (eds) *Forest Ecosystems, Forest Management, and the Global Carbon Cycle*, Proceedings of the NATO Advanced Research Workshop 'The Role of Global Forest Ecosystems and Forest Resource Management in the Global Cycle', held in Banff, Canada, 12–16 September, 1994, Springer, Berlin, Heidelberg

Forrester, D. I., Bauhus, J. and Khanna, P. K. (2004) 'Growth dynamics in a mixed-species plantation of *Eucalyptus globulus* and *Acacia mearnsii*', *Forest Ecology and Management*, no 193, pp81–95

Forrester, D. I., Bauhus, J., Cowie, A. L. and Vanclay, J. K. (2006) 'Mixed-species plantations of *Eucalyptus* with nitrogen-fixing trees: A review', *Forest Ecology and Management*, no 233, pp211–230

Forrester, D. I., Bauhus, J., Cowie, A. L., Mitchell, P. A. and Brockwell, J. (2007) 'Productivity of three young mixed-species plantations containing N_2-fixing *Acacia* and non-N_2-fixing *Eucalyptus* and *Pinus* trees in Southeastern Australia', *Forest Science*, no 53, pp426–434

Franklin, R. M. (2001) 'Permanent firebreaks: A useful land management tool', *Forest Landowner*, no 60, pp53–54

Freibauer, A. (2002) 'Biogenic greenhouse gas emissions from agriculture in Europe – quantification and mitigation', Institut für Landwirtschaftliche Betriebslehre, Universität Hohenheim

Freibauer, A., Böttcher, H., Scholz, Y., Gitz, V., Ciais, P., Mund, M., Wutzler, T. and Schulze, E. D. (submitted) 'Setting priorities for land management to mitigate climate change', *Climatic Change*

Friedlingstein, P., Dufresne, J. L., Cox, P. M. and Rayne, P. (2003) 'How positive is the feedback between climate change and the carbon cycle?', *Tellus Series B–Chemical & Physical Meteorology*, no 55, pp692–700

Fürstenau, C., Badeck, F., Lasch, P., Lexer, M., Lindner, M., Mohr, P. and Suckow, F. (2007) 'Multiple-use forest management in consideration of climate change and the interests of stakeholder groups', *European Journal of Forest Research*, no 126, pp225–239

Garcia-Gonzalo, J., Peltola, H., Zubizarreta Gerendiain, A. and Kellomäki, S. (2007) 'Impacts of forest landscape structure and management on timber production and carbon stocks in the boreal forest ecosystem under changing climate', *Forest Ecology and Management*, no 241, pp243–257

Gardiner, B. A. and Quine, C. P. (2000) 'Management of forests to reduce the risk of abiotic damage – a review with particular reference to the effects of strong winds', *Forest Ecology and Management*, no 135, pp261–277

Giardina, C. P. and Ryan, M. G. (2000) 'Evidence that decomposition rates of organic carbon in mineral soil do not vary with temperature', *Nature*, no 404, pp858–861

Giardina, C. P., Ryan, M. G., Hubbard, R. M. and Binkley, D. (2001) 'Tree species and soil textural controls on carbon and nitrogen mineralization rates', *Soil Science Society of America Journal*, no 65, pp1272–1279

Gonçalves, J. L. M., Wichert, M. C. P., Gava, J. L., Masetto, A. V., Arthur, J. C. Jr, Serrano, M. I. P. and Mello, S. L. M. (2007) 'Soil fertility and growth of *Eucalyptus grandis* in Brazil under different residue management practices', *Southern Hemisphere Forestry Journal*, no 69, pp95–102

González, J. R. (2006) 'Integrating fire risk into forest planning', Faculty of Forestry. University of Joensuu, Joensuu

Guo, L. B. and Gifford, R. M. (2002) 'Soil carbon stocks and land use change: A meta analysis', *Global Change Biology*, no 8, pp345–360

Gustavsson, L., Madlener, R., Hoen, H. F., Jungmeier, G., Karjalainen, T., Klöhn, S., Mahapatra, K., Pohjola, J., Solberg, B. and Spelter, H. (2006) 'The role of wood material for greenhouse gas mitigation', *Mitigation and Adaptation Strategies for Global Change*, no 11, pp1097–1127

Hansen, J., Ruedy, R., Sato, M. and Lo, K. (2006) 'GISS surface temperature analysis. Global temperature trends: 2005 summation', NASA Goddard Institute for Space Studies and Columbia University Earth Institute, http://data.giss.nasa.gov/gistemp/2005/

Harmon, M. E. and Marks, B. (2002) 'Effects of silvicultural practices on carbon stores in Douglas-fir–western hemlock forests in the Pacific Northwest, USA: results from a simulation model', *Canadian Journal of Forest Research–Journal Canadien de la Recherche Forestiere*, no 32, pp863–877

Harmon, M. E., Ferrell, W. K. and Franklin, J. F. (1990) 'Effects on carbon storage of conversion of old-growth forests to young forests', *Science*, no 247, pp699–702

Hirsch, K., Kafka, V., Tymstra, C., McAlpine, R., Hawkes, B., Stegehuis, H., Quintilio, S., Gauthier, S. and Peck, K. (2001) 'Fire-smart forest management: A pragmatic approach to sustainable forest management in fire-dominated ecosystems', *Forestry Chronicle*, no 77, pp357–363

Höhne, N. (2006) *What is Next After the Kyoto Protocol: Assessment of Options for International Climate Policy, Post 2012*, Purdue University Press, West Lafayette, IN

Houghton, R. A. (2003) 'Why are estimates of the terrestrial carbon balance so different?' *Global Change Biology*, no 9, pp500–509

Houghton, R. A., Hackler, J. L. and Lawrence, K. T. (2000) 'Changes in terrestrial carbon storage in the United States. 2: The role of fire and fire management', *Global Ecology and Biogeography*, no 9, pp145–170

Huston, M. A. and Marland, G. (2003) 'Carbon management and biodiversity', *Journal of Environmental Management*, no 67, pp77–86

IPCC (2001a) *Climate Change 2001: Mitigation. Contribution of Working Group III to the Third Assessment Report of the Intergovernmental Panel on Climate Change*, IPCC, Cambridge University Press, Cambridge

IPCC (2001b) *Climate Change 2001: The Scientific Basis. Contribution of Working Group I to the Third Assessment Report of the Intergovernmental Panel on Climate Change*, IPCC, Cambridge University Press, Cambridge

IPCC (2007) *Climate Change 2007: The Physical Science Basis. Summary for Policymakers. Contribution of Working Group I to the Fourth Assessment Report*, Intergovernmental Panel on Climate Change, Paris

Jactel, H., Brockerhoff, E. and Duelli, P. (2005) 'A test of the biodiversity stability theory: Meta-analysis of tree species diversity effects on insect pest infestation, and re-examination of responsible factors', in M. Scherer-Lorenzen, C. Körner and E.-D. Schulze, *Forest Diversity and Function: Temperate and Boreal Systems*, Springer, Berlin

Jandl, R., Lindner, M., Vesterdal, L., Bauwens, B., Baritz, R., Hagedorn, F., Johnson, D. W., Minkkinen, K. and Byrne, K. A. (2007) 'How strongly can forest management influence soil carbon sequestration?', *Geoderma*, no 137, pp253–268

Johnson, D. W. (1992) 'Effects of forest management on soil carbon storage', *Water, Air, and Soil Pollution*, no 64, pp83–120

Johnson, D. W. and Curtis, P. S. (2001) 'Effects of forest management on soil C and N storage: Meta analysis', *Forest Ecology and Management*, no 140, pp227–238

Johnson, D. W., Knoepp, J. D., Swank, W. T., Shan, J., Morris, L. A., Van Lear, D. H. and Kapeluck, P. R. (2002) 'Effects of forest management on soil carbon: Results of some long-term resampling studies', *Environmental Pollution*, no 116, pp201–208

Kahle, H. P., Karjalainen, T., Schuck, A., Ågren, G. I., Kellomäki, S., Mellert, K. H., Prietzel, J., Rehfuess, K. E. and Spiecker, H. (eds) (2008) *Causes and Consequences of Forest Growth Trends in Europe – Results of the RECOGNITION Project*, EFI Research Report, Brill Leiden, Boston, Köln.

Kaipainen, T., Liski, J., Pussinen, A. and Karjalainen, T. (2004) 'Managing carbon sinks by changing rotation length in European forests', *Environmental Science & Policy*, no 7, pp205–219

Karjalainen, T. (1996) 'Model computations on sequestration of carbon in managed forests and wood products under changing climatic conditions in Finland', *Journal of Environmental Management*, no 47, pp311–328

Kirschbaum, M. U. F. (2006) 'The temperature dependence of organic-matter decomposition – still a topic of debate', *Soil Biology and Biochemistry*, no 38, pp2510–2518

Knohl, A., Schulze, E. D., Kolle, O. and Buchmann, N. (2003) 'Large carbon uptake by an unmanaged 250-year-old deciduous forest in Central Germany', *Agricultural and Forest Meteorology*, no 118, pp151–167

Körner, C. (2003) 'Slow in, rapid out – Carbon flux studies and Kyoto targets', *Science*, no 300, pp1242–1243

Kurz, W. A. and Apps, M. J. (1999) 'A 70-year retrospective analysis of carbon fluxes in the Canadian forest sector', *Ecological Applications*, no 9, pp526–547

Kurz, W. A., Beukema, S. J. and Apps, M. J. (1998) 'Carbon budget implications of the transition from natural to managed disturbance regimes in forest landscapes', *Mitigation and Adaptation Strategies for Global Change*, no 2, pp405–421

Laiho, R., Sanchez, F., Tiarks, A., Dougherty, P. M. and Trettin, C. C. (2003) 'Impacts of intensive forestry on early rotation trends in site carbon pools in the southeastern US', *Forest Ecology and Management*, no 174, pp177–189

Liski, J., Pussinen, A., Pingoud, K., Makipaa, R. and Karjalainen, T. (2001) 'Which rotation length is favourable to carbon sequestration?', *Canadian Journal of Forest Research*, no 31, pp2004–2013

Magnani, F., Mencuccini, M., Borghetti, M., Berbigier, P., Berninger, F., Delzon, S., Grelle, A., Hari, P., Jarvis, P. G., Kolari, P., Kowalski, A. S., Lankreijer, H., Law, B. E., Lindroth, A., Loustau, D., Manca, G., Moncrieff, J. B., Rayment, M., Tedeschi, V., Valentini, R. and Grace, J. (2007) 'The human footprint in the carbon cycle of temperate and boreal forests', *Nature*, no 447, pp849–851

Malhi, Y., Meir, P. and Brown, S. (2002) 'Forests, carbon and global climate', *Philosophical Transactions of the Royal Society of London Series A – Mathematical Physical and Engineering Sciences*, no 360, pp1567–1591

Marland, G. and Marland, S. (1992) 'Should we store carbon in trees?' *Water, Air, and Soil Pollution*, no 64, pp181–195

Marland, G. and Schlamadinger, B. (1997) 'Forests for carbon sequestration or fossil fuel substitution? A sensitivity analysis', *Biomass and Bioenergy*, no 13, pp389–397

Masera, O. R., Garza-Caligaris, J. F., Kanninen, M., Karjalainen, T., Liski, J., Nabuurs, G. J., Pussinen, A., de Jong, B. H. J. and Mohren, G. M. J. (2003) 'Modeling carbon sequestration in afforestation, agroforestry and forest management projects: The CO2FIX V.2 approach', *Ecological Modelling*, no 164, pp177–199

MEA (Millennium Ecosystem Assessment) (2005) *Ecosystems and Human Well-being: Scenarios. Findings of the Scenarios Working Group*, Island Press, Washington, DC

Mead, D. J. (2005) 'Opportunities for improving plantation productivity. How much? How quickly? How realistic?' *Biomass and Bioenergy*, no 28, pp249–266

Merino, A., Balboa, M. A., Rodriguez Soalleiro, R. and Gonzalez, J. G. A. (2005) 'Nutrient exports under different harvesting regimes in fast-growing forest plantations in southern Europe', *Forest Ecology and Management*, no 207, pp325–339

Mund, M. and Schulze, E. D. (2006) 'Impacts of forest management on the carbon budget of European beech (*Fagus sylvatica*) forests', *Allgemeine Forst Und Jagdzeitung*, no 177, pp47–63

Nabuurs, G. J., Dolman, A. J., Verkaik, E., Kuikman, P. J., Diepen, C. A. v., Whitmore, A. P., Daamen, W. P., Oenema, O., Kabat, P. and Mohren, G. M. J. (2000) 'Article 3.3 and 3.4 of the Kyoto Protocol: Consequences for industrialized countries' commitment, the monitoring needs, and possible side effects', *Environmental Science & Policy*, no 3, pp123–134

Nabuurs, G. J., Masera, O., Andrasko, K., Benitez-Ponce, P., Boer, R., Dutschke, M., Elsiddig, E., Ford-Robertson, J., Frumhoff, P., Karjalainen, T., Krankina, O., Kurz, W., Matsumoto, M., Oyhantcabal, W., Ravindranath, N. H., Sanchez, X. and Zhang, M. J. S. (2007) 'Forestry', in Metz, B., Davidson, O. R., Bosch, P. R., Dave, R. and Meyer, L. A. (eds) *Climate Change 2007: Mitigation. Contribution of Working group III to the Fourth Assessment Report of the Intergovernmental Panel on Climate Change*, Cambridge University Press, Cambridge and New York

Narayan, C., Fernandes, P. M., van Brusselen, J. and Schuck, A. (2007) 'Potential for CO_2 emissions mitigation in Europe through prescribed burning in the context of the Kyoto Protocol', *Forest Ecology and Management*, no 251, pp164–173

Nilsson, L. O. (1997) 'Manipulation of conventional forest management practices to increase forest growth – Results from the Skogaby project', *Forest Ecology and Management*, no 91, pp53–60

Nilsson, S. and Schopfhauser, W. (1995) 'The carbon-sequestration potential of a global afforestation program', *Climatic Change*, no 30, pp267–293

Niu, X. Z. and Duiker, S. W. (2006) 'Carbon sequestration potential by afforestation of marginal agricultural land in the Midwestern US', *Forest Ecology and Management*, no 223, pp415–427

Nyland, R. D. (1996) *Silviculture: Concepts and Applications*, McGraw-Hill Companies, New York

O'Connell, A. M., Grove, T. S., Mendham, D. S. and Rance, S. J. (2004) 'Impact of harvest residue management on soil nitrogen dynamics in *Eucalyptus globulus* plantations in south western Australia', *Soil Biology and Biochemistry*, no 36, pp39–48

Palosuo, T., Peltoniemi, M., Mikhailov, A., Komarov, A., Faubert, P., Thürig, E. and Lindner, M. (2008) 'Projecting effects of intensified biomass extraction with alternative modelling approaches', *Forest Ecology and Management*, no 255, pp1423–1433.

Paul, K. I., Polglase, P. J., Nyakuengama, J. G. and Khanna, P. K. (2002) 'Change in soil carbon following afforestation', *Forest Ecology and Management*, no 168, pp241–257

Paul, K. I., Polglase, P. J. and Richards, G. P. (2003) 'Predicted change in soil carbon following afforestation or reforestation, and analysis of controlling factors by linking a C accounting model (CAMFor) to models of forest growth (3PG), litter decomposition (GENDEC) and soil C turnover (RothC)', *Forest Ecology and Management*, no 177, pp485–501

Pettersson, F. and Hogbom, L. (2004) 'Long-term growth effects following forest

nitrogen fertilization in *Pinus sylvestris* and *Picea abies* stands in Sweden',
Scandinavian Journal of Forest Research, no 19, pp339–347

Price, D. T., Halliwell, D. H., Apps, M. J., Kurz, W. A. and Curry, S. R. (1997)
'Comprehensive assessment of carbon stocks and fluxes in a Boreal-Cordilleran
forest management unit', *Canadian Journal of Forest Research–Revue Canadienne
De Recherche Forestiere*, no 27, pp2005–2016

Pussinen, A., Karjalainen, T., Makipaa, R., Valsta, L. and Kellomaki, S. (2002)
'Forest carbon sequestration and harvests in Scots pine stand under different
climate and nitrogen deposition scenarios', *Forest Ecology and Management*, no
158, pp103–115

Resh, S. C., Binkley, D. and Parrotta, J. A. (2002) 'Greater soil carbon sequestration
under nitrogen-fixing trees compared with *Eucalyptus* species', *Ecosystems*, no 5,
pp217–231

Rosenberg, O. and Jacobson, S. (2004) 'Effects of repeated slash removal in thinned
stands on soil chemistry and understorey vegetation', *Silva Fennica*, no 38,
pp133–142

Ryan, M. G., Binkley, D., Fownes, J. H., Giardina, C. P. and Senock, R. S. (2004) 'An
experimental test of the causes of forest growth decline with stand age', *Ecological
Monographs*, no 74, pp393–414

Saharjo, B. H. (1997) 'Fire protection and industrial plantation management in the
tropics', *Commonwealth Forestry Review*, no 76, pp203–205

Sampson, R. N. and Scholes, R. J. (2000) 'Additional human-induced activities –
Article 3.4', in R. T. Watson, I. R. Noble, B. Bolin, N. H. Ravindranath, D. J.
Verardo and D. J. Dokken, *Special Report on Land Use, Land-Use Change and
Forestry*, Cambridge University Press, Cambridge

Sanchez, F. G., Tiarks, A. E., Kranabetter, J. M., Page-Dumroese, D. S., Powers, R. F.,
Sanborn, P. T. and Chapman, W. K. (2006) 'Effects of organic matter removal and
soil compaction on fifth-year mineral soil carbon and nitrogen contents for sites
across the United States and Canada', *Canadian Journal of Forest Research*, no 36,
pp565–576

Sankaran, K. V., Chacko, K. C., Pandalai, R. C., Kumaraswamy, S., O'Connell, A. M.,
Grove, T. S. and Mendham, D. S. (2004) 'Improved productivity of eucalypt
plantations through site management practices in the monsoonal tropics: Kerala,
India', in E. K. S. Nambiar, J. Ranger, A. Tiarks and T. Toma (eds) *Site
Management and Productivity in Tropical Plantation Forests: Proceedings of
Workshops in Congo July 2001 and China February 2003*, CIFOR, Bogor,
Indonesia

Sathaye, J. and Bouille, D. (2001) 'Barriers, opportunities, and market potential of
technologies and practices', in B. Metz, O. Davidson, R. Stewart and J. Pan,
*Climate Change 2001: Mitigation. Contribution of Working Group III to the
Third Assessment Report of the Intergovernmental Panel on Climate Change*,
Cambridge University Press, Cambridge

Schär, C. and Jendritzky, G. (2004) 'Hot news from summer 2003', *Nature*, no 432,
pp559–560

Schimel, D. S., House, J. I., Hibbard, K. A., Bousquet, P., Ciais, P., Peylin, P.,
Braswell, B. H., Apps, M. J., Baker, D., Bondeau, A., Canadell, J., Churkina, G.,
Cramer, W., Denning, A. S., Field, C. B., Friedlingstein, P., Goodale, C., Heimann,
M., Houghton, R. A., Melillo, J. M., Moore, B., Murdiyarso, D., Noble, I., Pacala,
S. W., Prentice, I. C., Raupach, M. R., Rayner, P. J., Scholes, R. J., Steffen, W. L.
and Wirth, C. (2001) 'Recent patterns and mechanisms of carbon exchange by
terrestrial ecosystems', *Nature*, no 414, pp169–172

Schlamadinger, B. and Marland, G. (1994) 'Carbon implications of forest management strategies', in Apps, M. and Price, D. T. (eds) *Forest Ecosystems, Forest Management, and the Global Carbon Cycle*, Proceedings of the NATO Advanced Research Workshop 'The Role of Global Forest Ecosystems and Forest Resource Management in the Global Cycle', held in Banff, Canada, 12–16 September, Springer, Berlin, Heidelberg

Schulze, E. D. (2006) 'Biological control of the terrestrial carbon sink', *Biogeosciences*, no 3, pp147–166

Schulze, E. D., Wirth, C. and Heimann, M. (2000) 'Climate change – managing forests after Kyoto', *Science*, no 289, pp2058–2059

Seely, B., Welham, C. and Kimmins, H. (2002) 'Carbon sequestration in a boreal forest ecosystem: Results from the ecosystem simulation model, FORECAST', *Forest Ecology and Management*, no 169, pp123–135

Smith, K. A. and Conen, F. (2004) 'Impacts of land management on fluxes of trace greenhouse gases', *Soil Use and Management*, no 20, pp255–263

Spiecker, H. (1999) 'Overview of recent growth trends in European forests', *Water, Air, and Soil Pollution*, no 116, pp33–46

Stromgren, M. and Linder, S. (2002) 'Effects of nutrition and soil warming on stemwood production in a boreal Norway spruce stand', *Global Change Biology*, no 8, pp1195–1204

Thornley, J. H. M. and Cannell, M. G. R. (2000) 'Managing forests for wood yield and carbon storage: A theoretical study', *Tree Physiology*, no 20, pp477–484

Thuille, A. and Schulze, E. D. (2006) 'Carbon dynamics in successional and afforested spruce stands in Thuringia and the Alps', *Global Change Biology*, no 12, pp325–342

Tiarks, A. and Ranger, J. (2006) 'Soil Properties in tropical plantation forests: Evaluation and effects of site management: A summary', in E. K. S. Nambiar (ed.) *Site Management and Productivity in Tropical Plantation Forests*, Proceedings of Workshops in Piracicaba (Brazil) 22–26 November 2004 and Bogor (Indonesia) 6–9 November 2006, CIFOR, Bogor, Indonesia

UNECE/FAO (2000) *Forest Resources of Europe, CIS, North America, Australia, Japan and New Zealand, Contribution to the Global Forest Resources Assessment 2000*, United Nations, New York and Geneva

UNECE (2005) *European Forest Sector Outlook Study*, United Nations Economic Commission for Europe/Food and Agriculture Organization of the United Nations, Geneva

UNECE (2007) *Forest Products Annual Market Review, 2006–2007*, United Nations Economic Commission for Europe, Timber Section, Geneva, www.unece.org/timber/docs/fpama/2007/fpamr2007.htm

UNFCCC (1997) *Kyoto Protocol to the United Nations Framework Convention on Climate Change*, http://unfccc.int/resource/docs/convkp/kpeng.pdf

Verchot, L. V., Van Noordwijk, M., Kandji, S., Tomich, T., Ong, C., Albrecht, A., Mackensen, J., Bantilan, C., Anupama, K. V. and Palm, C. (2007) 'Climate change: Linking adaptation and mitigation through agroforestry', *Mitigation and Adaptation Strategies for Global Change*, no 12, pp901–918

Vesterdal, L. and Raulund-Rasmussen, K. (1998) 'Forest floor chemistry under seven tree species along a soil fertility gradient', *Canadian Journal of Forest Research*, no 28, pp1636–1647

Vieira, S., Trumbore, S., Camargo, P. B., Selhorst, D., Chambers, J. Q., Higuchi, N. and Martinelli, L. A. (2005) 'Slow growth rates of Amazonian trees: Consequences for carbon cycling', *Proceedings of the National Academy of Sciences of the United States of America*, no 102, pp18502–18507

Volney, W. J. A. and Fleming, R. A. (2000) 'Climate change and impacts of boreal forest insects', *Agriculture Ecosystems & Environment*, no 82, pp283–294

WBGU (1998) *The Accounting of Biological Sinks and Sources Under the Kyoto Protocol – A Step Forwards or Backwards for Global Environmental Protection?*, Wissenschaftlicher Beirat der Bundesregierung Globale Umweltveränderungen (WBGU), Berlin

West, T. O. and Marland, G. (2003) 'Net carbon flux from agriculture: Carbon emissions, carbon sequestration, crop yield, and land-use change', *Biogeochemistry*, no 63, pp73–83

Wilkie, M. L., Abdel-Nour, H., Carneiro, C. M., Durst, P., Kneeland, D., Kone, P. D., Prins, C. F. L., Brown, C. and Frisk, T. (2003) 'Forest area covered by management plans: Global status and trends', *Unasylva*, no 54, pp35–36

Williams, D. W. and Liebhold, A. M. (2002) 'Climate change and the outbreak ranges of two North American bark beetles', *Agricultural and Forest Entomology*, no 4, pp87–99

Woodbury, P. B., Heath, L. S. and Smith, J. E. (2006) 'Land use change effects on forest carbon cycling throughout the southern United States', *Journal of Environmental Quality*, no 35, pp1348–1363

Yanai, R. D., Currie, W. S. and Goodale, C. L. (2003) 'Soil carbon dynamics after forest harvest: An ecosystem paradigm reconsidered', *Ecosystems*, no 6, pp197–212

Zerva, A. and Mencuccini, M. (2005) 'Carbon stock changes in a peaty gley soil profile after afforestation with Sitka spruce (*Picea sitchensis*)', *Annals of Forest Science*, no 62, pp873–880

4

Planted forests and water

Rodney J. Keenan and Albert I. J. M. Van Dijk

Introduction

This chapter provides an overview of the effects of planted forests on water resources. Planted forests include both intensively managed forest plantations and more extensively managed planted forests (*sensu* FAO, 2004) that have been established for wood production, water and soil protection, landscape restoration, biodiversity conservation or other purposes. In most cases, the general impacts on water resources of different types of planted forests will be similar, with differences resulting from tree species selection or management activities.

Global forest cover is in a continuing state of change. Human actions are resulting in deforestation and forest loss, and forest expansion is occurring due to afforestation, forest plantation establishment and the abandonment of agricultural land and natural forest regeneration. The global plantation estate grew by 42 per cent between 1990 and 2005, to a total of 139.1 million hectares (Mha), or about 3.5 per cent of total forest cover (FAO, 2005a). About half of this expansion involved conversion from native forest to plantation. Most deforestation and conversion from natural to plantation forest (accounting for about half of total plantation increase) occurred in humid tropical regions. Conversion of agricultural land to plantation forests largely occurred in eastern China and southern Europe. However, expansion also occurred in several other industrialized and industrializing countries (FAO, 2005a).

Maintaining intact forests, expanding the forest area and restoring forests on degraded lands have been promoted in many policy discussions because of their perceived benefits for water quality and, often, water yield. Increased forest plantations are in many cases being actively supported by governments for economic development or for land restoration purposes, through the provision of subsidies or loans, favourable regulation or other supporting policy mechanisms (Enters and Durst, 2004). Plantation forests can provide economic, social and environmental benefits (Gerrand et al, 2003) including watershed protection benefits. The Kyoto Protocol provided for the establishment of forest in previously cleared areas as a climate change mitigation measure and it has been actively promoted for this purpose (Jackson et al, 2005) as well as for bioenergy measures (Field et al, 2007). However, due to a range of impediments, relatively few forest-based projects are incorporated into the Clean Development Mechanism (CDM) of the Protocol.

While the value of forested, protected areas in providing water quality benefits has been widely promoted (Dudley and Stolton, 2003), it is also recognized that global water resources are coming under increasing pressure to meet the needs of an expanding population, an increasing dependence of intensive agriculture on irrigation water and increasing demand for hydro-electric energy generation. Where there has been extensive forest clearing there is often more water in the landscape and water use by communities or industries is often based on this increased water availability. Allocation of these water resources has often been in excess of that available and, consequently, significant industry and community adjustment is required to match demand to supply. Increased forest cover can further compound this adjustment process and, more often than not, the poorest and the environment suffer the greatest consequences of reduced water availability (Dye and Versfeld, 2007).

In many water-limited regions, increases in irrigated land have helped sustain a growing population. It is estimated that 90 per cent of the global consumptive use of extracted water resources is for irrigation (Shiklomanov, 1997) and further agricultural growth is increasingly constrained by water (e.g. Oki and Kanae, 2006). By 2050, 30–40 per cent more fresh water could be used by agriculture than is used today (De Fraiture et al, 2007).

Changing climate (most likely as a result of human influence) is also likely to result in lower average rainfall in some regions, altered seasonality of rainfall and reductions in water supply through increased evaporation. Combined with a growing population, this will place increasing demands on water resources and the need to utilize water resources for food production more efficiently (Bossio et al, 2009).

Debate over the hydrological benefits or impacts of afforestation has occurred for some time (Dye and Versfeld, 2007). The weight of scientific evidence over the last 30 years indicates that large-scale afforestation can impact on water resources. Increased forest cover results in reduced streamflow and groundwater inputs, and increased no- or low-flow days, although this varies considerably between regions and ecosystems and with the scale and type of reforestation.

Potential impact on water resources has contributed to a wider debate about the merits of intensive plantation forestry (e.g. Cossalter and Pye-Smith, 2003; Jackson et al, 2005). Some forest industry sectors consider that plantation forestry is being 'singled out' for particular policy attention and this is more severe when plantation development is conducted directly by the state or is supported, directly or indirectly, through taxation or other incentives.

Forests and rainfall

The impact of forests on larger-scale regional or global hydrological cycling has been a matter of scientific conjecture for some time. Forests affect the global heat balance through their direct influence on radiation balance, and afforestation of agricultural land can lead to a lower albedo and greater absorption of radiant energy, an effect which is most pronounced in snow-prone areas (Pielke et al, 2006). Changes in atmospheric vapour content after afforestation or deforestation may influence radiation balance, although this effect will vary between climate zones and seasons. Greater canopy roughness can lead to more efficient mixing of the lower atmosphere above forests and lower absorption of radiant energy than over agricultural land, although these changes depend on soil water availability. Such effects might result in changes in regional circulation and weather patterns, including rainfall generation. Afforestation can affect near-surface temperatures and vapour flux into the atmosphere through changes in albedo and evapotranspiration (Bruijnzeel, 2004; Van der Molen et al, 2006; Pielke et al, 2006). Large-scale removal of forest could reduce onshore wind strength, cloud levels and convection patterns, particularly over islands and coastal areas in the humid tropics, which may be important for the global redistribution of atmospheric moisture (Van der Molen et al, 2006) and lead to reduced rainfall (Pielke et al, 2006).

At the continental scale, the relationship between annual precipitation and distance from the ocean may change after large-scale forest cover changes. Makarieva et al (2009) argued that a 'biotic forest pump' would be able to maintain constant or even increase inland precipitation, but Meesters et al (2009) demonstrated that their argument was based on incorrect physics and was unsustainable. Pitman et al (2004) suggested that reduced surface roughness following forest and woodland clearing largely explained reduced rainfall patterns in south-west Western Australia and that rainfall could be returned to the long-term average through large-scale reforestation in this region. The scale of such afforestation would need to be of the order of hundreds of thousands of hectares to achieve such results. Partial afforestation would create more heterogeneous landscapes, potentially with quite different outcomes (Pielke et al, 2006).

Flood control and landslides

There has been long-term debate about the role of forests in reducing the frequency, intensity and impacts of floods. Much of this debate has become embedded in other political, institutional or policy agendas (Calder and Aylward, 2006; Van Dijk and Keenan, 2007; Bradshaw et al, 2009) and it is difficult to separate science from alternative positions relating to maintaining existing land-use patterns or encouraging land-use change. Most hydrological investigations of forests and water have focused on the impacts on low flows and water yield and there has been little comprehensive analysis of the impact of land cover change on peak flows. Hewlett (1982) demonstrated that the presence or absence of forest did not appreciably influence the magnitude of the largest flow events. Peak flows after medium to high rainfall events may be reduced after afforestation in smaller catchments (e.g. Bruijnzeel, 2004; Scott et al, 2005; Waterloo et al, 2007) and localized 'flash floods' associated with short but intensive rainfall events may also be reduced. At larger catchment or basin scales there is no strong empirical evidence or theoretical argument to expect a reduction of flooding in large basins (FAO, 2005b).

In a recent international analysis, Bradshaw et al (2007) suggested that flood frequency is negatively correlated with the amount of remaining natural forest within countries and positively correlated with natural forest area loss. Loss of forests may increase or exacerbate the number of flood-related disasters, negatively impact millions of poor people and inflict trillions of dollars worth of damage on disadvantaged economies over the coming decades. Reforestation, therefore, might help to reduce the frequency and severity of flood-related catastrophes. Laurance (2007) provided strong support for their position. An alternative analysis showed the frequency of flood events impacting on people was much more strongly correlated to population density (Van Dijk et al, 2008). This is not surprising for various reasons, including the fact that the data used were based on media reporting. After considering this relationship, there was no residual correlation between flood impacts and forest cover.

Flooding is a result of the combination of rainfall (usually a high-intensity event but of varying spatial extent and duration), topography and land use that combine to determine flows in particular parts of a river system. Conversion of land to non-forest use can result in reduced soil infiltration as a consequence of logging, agricultural or urban land use, or road construction and altered drainage, increased sediment generation and loss of channel capacity and overbank flows. Therefore, it is difficult, if not impossible, to single out tree clearance as a cause of flooding, except in controlled, small-scale experiments. Factors such as soils, geology, catchment and river morphology, and antecedent conditions (e.g. catchment wetness) affect the generation of floods. Overall, there appears to be little empirical hydrological evidence or theoretical argument that the removal of trees is likely to exacerbate severe flooding or that large-scale afforestation will have major flood mitigation benefits (Van Dijk et al, 2008). This brings into question the rationale for large-scale afforestation programmes aimed at achieving such outcomes (Calder, 2007).

Afforestation also appears to have little impact on large, deep-seated landslides, which usually appear to be triggered by the development of a water table above the bedrock interface during prolonged periods of high rainfall (Bruijnzeel, 2004; Sidle et al, 2006). Well-established tree cover may reduce the risk of shallow landslides, as tree root systems can provide structural strength to several metres depth. Conversely, however, the presence of trees itself can lead to log jams in streams and so contribute to local flooding (Sidle et al, 2006). The net outcome of these processes is difficult to predict.

Water yield

It has been demonstrated through a wide range of scientific investigations that trees use more water than shallow-rooted forms of vegetation such as crops, pastures, grasslands or shrublands, and that catchments covered by forests produce less surface run-off, groundwater recharge and streamflow. This evidence has been generated at varying scales: at the tree level using sap flow measurements (Wullschleger et al, 1998; Almeida et al, 2007), at the site level using water balance or atmospheric water flux studies (e.g. Wilson et al, 2001; Petheram et al, 2002; Dye and Versfeld, 2007), at the catchment scale using controlled and paired catchment studies (Waterloo et al, 2007; Buytaert et al, 2007), and through statistical analyses of streamflow data from multiple catchments (Bosch and Hewlett, 1982; Zhang et al, 2004; Scott et al, 2005; Van Dijk et al, 2007).

Trees and forests have taller, rougher and denser canopies than other vegetation types, generally resulting in greater interception losses (evaporation of rainfall from the canopy; Van Dijk and Bruijnzeel, 2001). Interception losses are usually smaller for deciduous forests than for needle-leaf forests (typically ca. 20 per cent vs. 30 per cent of rainfall; e.g. Roberts, 1999), particularly where most rain falls in winter. Hanging and waxy leaves, such as those of many eucalypt species, reduce the amount of rain retained in the canopy and subsequently evaporated.

The impact of afforestation on water resources needs to be evaluated against a benchmark condition, either the current mix of agricultural land uses or the natural vegetation that existed before clearing. The latter should take into account the mix of stand development or successional vegetation stages that might have existed prior to clearing. The former is most appropriate when evaluating the consequences of water resource availability for irrigation or industry and may be more appropriate for landscape restoration projects (Van Dijk and Keenan, 2007).

The effect of forest cover on run-off and water yield increases with annual rainfall. In a meta-analysis of catchment studies, Zhang et al (2004) estimated that absolute reductions in water use due to forest cover averaged about 3 megalitres per hectare in regions with 1500mm per year rainfall and 2 megalitres per hectare when mean annual rainfall was 1000mm. The difference was negligible in catchments with less than 500mm per year. However, the proportional reduction in water use as a result of afforestation is higher in seasonally dry climates, where the deeper roots of trees can provide access to a greater depth of soil water, and sometimes groundwater (Benyon et al, 2006), and trees can therefore sustain greater transpiration rates through dry periods (e.g. Nepstad et al, 1994; Dye and Versfeld, 2007).

Fast-growing species and dense forests generally have greater water use than slower-growing species or sparse forest (e.g. Hatton et al, 1998). Differences between tree species in average transpiration and transpiration patterns over time can arise from differences in biomass allocation (i.e. root, leaf or structural growth), root distribution, hydraulic architecture and stomatal response. Most forest plantation species have been selected for rapid early growth with annual stem increment peaking at a relatively young age and this is reflected in water use. Studies in plantation forests (Farley et al, 2005; Almeida et al, 2007) and native forests after fire or other disturbance (Kuczera, 1987; Vertessy et al, 2001; Giambelluca, 2002) confirm that both leaf area and water use initially increase quickly and then gradually decrease with age. For plantations managed for timber production, the impact will depend on the point in development at which the stand is harvested, whereas forests established for other purposes may be left to mature and the water use will potentially decline over time.

Comparative studies in South Africa suggest that eucalypt plantations use more water than pine plantations (Scott et al, 2005), whereas the reverse has been observed in Australia (Putuhena and Cordery, 2000). This may well be related to endemic insect herbivores and other pests that reduce leaf area and water use in Australia (Cornish and Vertessy, 2001). In a long-term paired catchment study in southeastern Australia, Bren and Hopmans (2007) showed that water use by a radiata pine plantation was generally lower than a nearby native eucalypt forest, except when the plantation was very young and during very dry years. Timber volume growth is generally more rapid in the radiata pine plantation and therefore water-use efficiency (in terms of timber production per unit of water transpired) is likely to be higher in that species.

Scale is an important factor. Development of models to predict the impacts of afforestation on water yield for larger catchments are often based on experimental results from smaller, steeper and more homogeneous catchments (Van Dijk et al, 2007). Larger catchments are likely to have different topography with larger, flatter and less immediately responsive areas, and a high variability in vegetation and land use, and in structures to conserve, collect and store water (e.g. farm bunds, contour tilling, farm dams) or weirs and other water controlling devices. This calls for caution when predicting the impact of afforestation on the basis of knowledge from small experimental studies under controlled, 'idealized' conditions. For example, Van Dijk et al (2007) estimated that reduction in total surface water resources across the Murray Darling basin due to future plantation expansion would be <0.3 per cent by 2020, and plantation expansion was not predicted to alter the flow regime in large catchments, although it may do so in catchments smaller than 2000km^2. Previous modelling had predicted much larger impacts, albeit mainly because forestry expansion scenarios were more extreme.

Soil properties exert an important influence on the water balance of agricultural catchments, and this introduces further variation. Globally, many agricultural areas have degraded and exposed soils with reduced infiltration and soil water-holding capacity, leading to increased surface run-off generation and erosion (Bruijnzeel, 2004). Lower soil disturbance or compaction in forests, the presence of a litter layer and often greater porosity under plantations may lead to higher infiltration and stored soil water. Afforestation in such degraded landscapes can lead to an improvement in soil infiltrability (e.g. Ilstedt et al, 2007) if the soils are sufficiently resilient. Depending on the catchment topography and geology this can ultimately flow to streams or be transpired later in the growing season (Bruijnzeel, 2004; Ilstedt et al, 2007). Where soil hydrological function is recovered, surface run-off generation, storm water damage and pollutant mobilization may be reduced, and groundwater recharge may increase if increased infiltration outweighs increased water use by tree vegetation. This could enhance low flows if the groundwater response time is such that these volumes are subsequently released during low-flow periods. Vanclay (2008) also suggests that most plantation establishment involves actively modifying the soil to reduce run-off (such as contour ploughing) and improve infiltration, and that these changes may reduce the effects of afforestation on base or peak flows.

Consequently, the idea that afforestation may 'make springs and rivers flow again' is possible but probably the exception rather than the rule. It has been demonstrated in selected cases, for example in China (McVicar et al, 2007) and Indonesia (Bruijnzeel, 2004; Scott et al, 2005; Chandler, 2006). In most other cases, however, low flows will be reduced after afforestation. This is more likely to be a concern to extractive users than to river ecology because, for most catchments, tree cover will have existed historically and the river ecology will have adapted to low flows. Exceptions where afforestation may cause a hydrological change away from pre-development conditions include natural grassland systems, e.g. in South Africa (Dye and Versfeld, 2007) and the high altitude parano grassland in Ecuador (Buytaert et al, 2007).

For agricultural irrigation or domestic water supplies based on river extraction, concern will typically be over the extension of low-flow periods and the implications that this has for minimum flow security and river ecosystems. In water-limited environments, the relative reduction in low flows is usually greater than in average flows (see Brown et al, 2007; Dye and Versfeld, 2007). However, it is likely that some bias in the selection of experimental catchments may confound this interpretation (Van Dijk et al, 2007).

A further distinction needs to be made between the water 'used' by trees vs. the volume that will no longer be available further downstream. In many large basins in water-limited areas, run-off is generated in the wetter uplands, stored in large impoundments and then released to users at a considerable distance downstream (e.g. in India, China, South Africa and Australia). Considerable amounts of water can evaporate from the river and water distribution system (including riparian zones, floodplains, weirs, reservoirs and irrigation supply channels) or leak to groundwater systems that are not accessible to users. A reduction in the volume of upstream run-off generation (from afforestation or another cause) will therefore normally translate into a smaller volume reduction downstream. River flow models are needed to evaluate these processes. This may reveal some surprises, with some plantation establishment having little or no impact on downstream water availability (see Brown et al, 2007; CSIRO, 2008).

In the catchments of large reservoirs, the overall impact of afforestation will depend on the capacity to release water to downstream users as and when desired. This is determined by the overall decrease in inflow volumes over longer timescales (the actual scale depends on the comparative magnitudes of storage capacity, inflows and releases). This is particularly true in situations where most, or all of the resource is allocated and the need for water security is high (e.g. for water supply to cities and towns and perennial crop growers; Brown et al, 2007; CSIRO, 2008).

The overall impact on water yield will depend on the extent of the catchment planted and the spatial configuration of the trees. It is difficult to reliably detect the effects on run-off if the proportion of the catchment area reforested is below a certain threshold because rainfall and the hydrological effect of the catchment surface, soil profile and vegetation vary across catchments. The threshold is typically 15–20 per cent of area planted in smaller catchments, but it is lower in larger catchments (Bosch and Hewlett, 1982; Stednick, 1996; Zhang et al, 2006; but see Trimble et al, 1987). These are statistical findings that are presumably related to remaining unexplained variation and noise in the data rather than representing a real threshold. Still, they suggest that the impact from limited increases in a plantation estate at a larger catchment scale is small (see Van Dijk et al, 2007).The proportion of afforestation that will cause a measurable change in a larger catchment also depends on the extent of variation in rainfall and run-off over the area.

Managing impacts on water yield

Given the social and political concerns about reduced water yield associated with plantation and planted forest development, a range of options have been put forward to mitigate the impacts of afforestation on water yield. Soils tend to get deeper and accumulate more surface run-off, soil water and ground-water in lower sections of catchments. Vertessy et al (2003) suggested that, because of this, planting trees on upper slopes will potentially have much less impact on streamflow than planting the same area on lower slopes, at least in water-limited environments, although partial clearing experiments have not found any evidence for similar differences in the (more humid) Eastern USA (Hornbeck et al, 1993). Zoning of plantations away from streams and drainage lines is already commonplace in many countries (though usually for other reasons) and, depending on the alternative riparian vegetation, this may reduce the overall impact of trees on water resources. The same rationale has led to encouraging the removal of invasive riparian vegetation in the semi-arid south-western USA (Newman et al, 2006). Vanclay (2008) proposed designing the plantation canopy to maximize decoupling by managing canopy roughness and air turbulence induced by abrupt plantation edges and fire beaks. He suggested that mixed-species plantings may have a canopy structure that is less coupled to the atmosphere, leading to reduced transpiration (see also Forrester, 2007).

Water quality

Afforestation and catchment management has long been promoted to reduce soil erosion, nutrient and sediment input to streams. Surface run-off is a major driver of soil erosion and river pollution, although river incision, mass wasting of hillsides or river banks and other forms of disturbance (such as fire) can result in large earth movement and volumes of sediment to streams and reservoirs (Bruijnzeel, 2004; Sidle et al, 2006). Afforestation can reduce surface run-off and soil erosion. Tree cover alone affords some protection by reducing the direct impact of rainfall, but maximum erosion protection is provided with the development of a litter layer, understorey growth and the surface roughness provided by tree roots. The application of fertilizers, and potential nutrient contribution to streams are considerably lower in plantation systems than in other forms of agricultural land use (e.g. May et al, 2009). Some forms of plantation management may result in higher nutrient input to streams during the harvesting phase (Nykvist et al, 1994; Malmer, 1996; Mackensen et al, 2003). Belts of trees can act as 'filter strips', intercepting and transpiring surface run-off and nutrients before they reach the stream (Ellis et al, 2006).

On the other hand, damage to the soils and drainage network during plantation establishment, harvesting and poor road network design can mobilize large volumes of sediment and offset the positive effects of greater soil protection and reduced surface run-off (Bruijnzeel, 2004; Waterloo et al, 2007).

Salinity is a major problem in many seasonally dry or semi-arid areas, particularly in geologically older landscapes on all continents where salts have had a long time to accumulate and the rainfall has not been high enough to flush these from the soil profile. Afforestation can minimize inputs of salt that negatively impact on water quality, soil properties and land suitability for agriculture. This occurs in two ways:

1 Dryland salinity often occurs after the clearing of woody vegetation; recharge is increased and the water table rises in lower parts of the landscape. Where salts are present deeper in the soil profile, these are mobilized and deposited close to the soil surface. The decreased run-off and streamflow associated with afforestation, discussed above, can lower the water table and reduce the extent of dryland salinity.
2 Salts can also be mobilized directly into streams with increased run-off after deforestation. Conversely, afforestation can lead to a reduction in stream salinity if it reduces inflows of more saline water (see Van Dijk and Keenan, 2007).

Salt or pollutant *concentrations* are often more of a problem for water users and stream health than total load. In this case, the interaction between reduced pollutant generation and reduced streamflow is important. Salinity concentrations in larger river systems can vary over time as salt load changes due to changes in the relative contributions of surface run-off and total streamflows. A whole-of-system analysis is necessary to evaluate outcomes further downstream, accounting for spatial and temporal patterns in run-off and pollutant generation, water losses from the system and any pollutant decay or deposition. In most regions, salts are located in specific parts of the landscape and the contribution of different catchments to stream salinity will vary widely. It is important to undertake effective survey and planning to locate afforestation for salinity mitigation in those parts of the landscape where the salt is being mobilized. Planting in catchments that contribute water with low salt concentrations could result in an increase in salt concentrations in major streams (Baker and Barson, 2006; Van Dijk et al, 2007). Thus, there are often trade-offs between decreasing water yield and increasing water quality associated with planted forest in catchments (see also McVicar et al, 2007).

Scenario and assessment tools

Given the challenges of direct observation of the impacts of planted forests on water yield or quality, and the long timescale between afforestation actions and catchment responses, a variety of analytical tools have been developed to assess the potential impact of afforestation on water. Models currently used to assess the impact of conversion of agricultural land to plantations are typically either directly derived from catchment data or from process simulation models.

Models derived from catchment data are generally based on statistical regression equations or simple concepts that explain part of the observed differences in long-term streamflow from forested and non-forested catchments (e.g. Bosch and Hewlett, 1982; Zhang et al, 2004). These have been integrated into more sophisticated decision support tools to analyse the potential impacts of afforestation on water properties in different parts of the world (McVicar et al, 2007; Van Dijk et al, 2007). While statistically robust, the predictive power of these models for specific applications is limited. Relationships commonly used do not account for differences in species, stocking, plantation age, other management factors (see Chapter 5) or the variety of conditions that might be found within catchments. One obstacle is the large amount of data required to derive a statistically significant relationship due to the often stronger variation caused by (unknown) climate and terrain factors, and sampling or other errors.

Mechanistic approaches (Chappell et al, 2007) may be more promising. These can account for differences in soil properties, rooting depth and vegetation characteristics (e.g. leaf area or surface conductance), and some include dynamic growth models that can simulate the effect of stocking density, soil fertility and management regime (e.g. Gallant et al, 2005). While useful for sensitivity studies, the accuracy and reliability of these models is also constrained by data availability, particularly for soil properties (see Scott et al, 2005; Ilstedt et al, 2007).

Payments for watershed services from planted forests

As indicated at the beginning of this chapter, there has been considerable debate in recent years regarding the merits of supporting afforestation programmes and the development of planted forests for catchment protection and watershed benefits. While the generally perceived view was that 'any tree, anywhere' is automatically a good thing for erosion protection and soil and water benefits, this notion is increasingly coming under question (Alexander and Campbell, 2003). Concerns probably arose first with the large-scale afforestation of natural grasslands in South Africa (Dye and Versfeld, 2007), but this has grown to a more widespread concern about the impacts of afforestation in regions where water resources are limited (Calder, 2007). Policy-makers are now often faced with a dilemma over whether to actively support forest regeneration and restoration programmes or not. Caps on the extent of catchments that can be afforested have been implemented in some jurisdictions (South Africa, South Australia) and are being considered at the national level in Australia. In other places, such as China and Central America, policies are actively supporting the establishment of planted forests for perceived hydrological benefits (Calder, 2007).

Various approaches can create greater public and land manager awareness of unpriced environmental assets resulting in behaviour changes leading to improved environmental outcomes (Pannell, 2008). Mechanisms for managing the hydrological impacts and benefits of afforestation, landscape restoration or catchment management programmes can include education or incentive arrangements in addition to these types of regulatory arrangements. Often a combination of all three is required to fully achieve policy objectives. Where regulation is applied, traditional ownership rights are changed through government action, and there is often a demand for compensation from existing owners and the costs of implementing this approach can outweigh the benefits (Aretino et al, 2001). Increasingly, governments are looking toward incentive programmes and market-based payment arrangements to encourage changes in behaviour and provide for efficient allocation of public funds, or to discriminate effectively between public and private benefits of forest restoration programmes and set up arrangements where the beneficiaries of the services provided by forests pay for their establishment and management (see Chapter 2). Economic incentives can consist of taxes, liability for damages, subsidies for 'better behaviour' or contracts for the purchase of improved environmental outcomes (Shortle and Horan, 2008). There are three potential sources of capital for investment in contracts: government, voluntary private sources and regulated private investment (Binning et al, 2002).

The most common type of market-based trading arrangements for environmental services or pollution control involve setting limits on the total amount of activity and issuing tradable permits such as for carbon dioxide emissions trading. These provide a basis for industry participants to determine the cheapest abatement options. These can be applied most effectively where there are relatively few known yet sufficient sources of emissions to form an effective market (US EPA, 2001).

Services provided by planted forests can be considered more in the class of 'non-point pollution control problem' (Shortle and Horan, 2008). Non-point sources of pollution, such as leaching and run-off of nutrients or chemicals from farm fields or urban streets do not have these characteristics. They are often unobservable, difficult or impossible to measure directly and the output cannot generally be ascribed to individuals or firms. These are most often controlled through regulation or education programmes. Measurement, enforcement and control costs for these types of pollution sources are costly. The most efficient approach economically is to focus control efforts on ambient pollution concentrations and the overall abatement costs of different measures. However, many agents that cause little or no problem might be subject to regulation. Options for reducing these types of transaction costs include (1) creating incentives for accurate self-reporting, (2) restricting incentive payments to activities that are easy to observe and highly correlated with ambient impacts or (3) shifting from monitoring many inputs from individual sources to a more easily observed alternative.

Perhaps the most celebrated case of watershed benefits of forest restoration was the decision by the City of New York to invest about US$1 billion in land protection and conservation practices to avoid spending US$4–6 billion on filtration and treatment plants (Perrot Maitre and Davis, 2001). Much of the effort focused on improving the management of existing forest and changing agricultural land-use practices, but there was some support provided for afforestation. Other examples of payment arrangements for forest watershed services include (Perrot Maitre and Davis, 2001):

1 In Costa Rica, a utility company pays into a fund that pays private upstream landholders to increase forest cover to provide regularity of water flow for hydroelectricity generation.
2 In the Cauca Valley, Colombia, associations of irrigators pay additional fees to a regional agency for land and forest activities to obtain a sufficient supply of water for crops.
3 In New South Wales, Australia, a farmers' cooperative buys 'transpiration credits' from the state forest management agency. The agency earns credits by reforesting upstream lands – a process expected to result in a reduction of water salinity downstream.

Ensuring purchasers get what they pay for is basic to any financial transaction. The scientific principles and tools described above can be used to develop improved understanding of cause and effect relationships between land cover and water quality or yield benefits and to measure how these elements change over time. The key challenges in relation to market-based approaches to watershed benefits are:

- Developing payment systems that address the often long timeframe between action and consequence in relation to watershed benefits. It may take at least a decade for any hydrological impact of afforestation to express itself. In regions with old geological structures and large regional flow systems these benefits may take hundreds of years to realize.
- Developing payment systems that address trade-offs with the production of other goods, benefits or services. The interaction between water quality and water yield is a particular problem for watershed managers. Intact forests may provide the best water quality but lower potential water yield. Interactions with the production of agricultural goods, timber and other services such as carbon sequestration or biodiversity conservation are also important (see Chapter 5). While there may be some 'win–win' combinations, there will often be trade-offs between different environmental services.
- Addressing the uncertainty associated with the outcome of land-use or land-cover change. This uncertainty can be due to: the lack of basic catchment data (such as geology, soils or climate); limited scientific understanding of the processes operating in a catchment or the effects of the land-use change; changes in future climate; variation in disturbance regimes such as fire, insect pests or diseases that have unintended hydrological consequences; or unforeseen changes in patterns of human settlement or land use in catchments over time that impact on the hydrological conditions in the catchment.

Conclusions

Understanding of the interactions between planted forests and water has improved considerably over the last 20–30 years. It is now clear that trees tend to use more water than shallow-rooted forms of vegetation such as crops, pastures, grasslands or shrublands and that catchments covered by forests produce less surface run-off, groundwater recharge and streamflow. Absolute impacts on streamflow are greatest in high rainfall (>1500mm per annum) catchments but the proportional reduction is higher in lower rainfall (500–700mm per annum) catchments. Afforestation often results in an increase in the number of low-flow or no-flow days. However, the spatial scale is important: except in smaller catchments (say <100ha), it is difficult to detect the impact of afforestation on hydrology if less than 20 per cent of the catchment is planted (Zhang et al, 2006). There are some differences between species in these effects, and impacts can be mitigated through plantation management activities such as belt and strip plantings, thinning and managing rotation lengths. However, in catchments with soils degraded by agricultural or other activities, afforestation may increase infiltration, water storage and streamflow.

The impact of afforestation and development of planted forests on minimizing the impact of flooding is likely to be limited to smaller catchments.

Planted forests may reduce the frequency of shallow earth movements and landslides but are not likely to impact on larger-scale debris flows and earth movement, which are primarily a result of the combination of extreme rainfall and catchment hydrology.

Assessing the impacts of planted forests on water yield and availability will depend on the nature of the catchments, impoundment and water management systems and losses due to evaporation and other impacts. Simplistic calculations based on forest evapotranspiration may give a misleading picture of the impacts of plantations on water availability downstream. More sophisticated decision support tools integrating a process-based understanding of forest physiology, geological and hydrological processes and climatic variables can provide more realistic assessments of the potential impact of planted forests on water.

Policy mechanisms to provide for better management of water and planted forests can include regulation, education and market-based incentive payment schemes. These arrangements will need to consider the long timeframes between actions and impacts, interactions with the production and payment of other goods and services such as food, timber, carbon and biodiversity conservation and the uncertainties in predicting the hydrological consequences of planted forest establishment.

References

Alexander, J. and Campbell, A. (2003) 'Plantations and sustainability science: The environmental and political settings', *Australian Forestry*, vol 66, pp12–19

Almeida, A. C., Soares, J. V., Landsberg, J. J. and Rezende, G. D. (2007) 'Growth and water balance of *Eucalyptus grandis* hybrids plantations in Brazil during a rotation for pulp production', *Forest Ecology Management,* vol 251, pp10–21

Andreassian, V. (2004) 'Waters and forests: From historical controversy to scientific debate', *Journal of Hydrology*, vol 291, pp1–27

Aretino, B., Holland, P., Matysek, A. and Peterson, D. (2001) *Cost Sharing for Biodiversity Conservation: A Conceptual Framework*, Productivity Commission, Melbourne

Baker, P. and Barson, M. (2006) *Sourcing the Salt: How to Find Out if Your Catchment is Contributing to a Salinity Problem*, Science for Decision Makers Brief, Bureau of Rural Sciences, Canberra, Australia

Benyon, R. G., Theiveyanathan, S. and Doody, T. M. (2006) 'Impacts of tree plantations on groundwater in south-eastern Australia *Australian Journal of Botany*, vol 54, pp181–192

Binning, C., Baker, B., Meharg, S., Cork, S. and Kearns, A. (2002) *Making Farm Forestry Pay – Markets for Ecosystem Services*, Rural Industries Research and Development Corporation Publication No 02/005, Canberra, Australia

Bosch, J. R. and Hewlett, J. D. (1982) 'A review of catchment experiments to determine the effect of vegetation change on water yield and evapotranspiration', *Journal of Hydrology*, vol 55, pp3–22

Bossio, D., Geheb, K. and Critchley, W. (2009) 'Managing water by managing land: Addressing land degradation to improve water productivity and rural livelihoods', *Agricultural Water Management*, vol 97, pp536–542

Bradshaw, C. J. A., Sodhi, N. S., Peh, K. S. H. and Brook, B. W. (2007) 'Global evidence that deforestation amplifies flood risk and severity in the developing world', *Global Change Biology*, vol 13, pp2379–2395

Bradshaw, C. J. A., Brook, B. W., Peh, K. S. H. and Sodhi, N. S. (2009) 'Flooding policy makers with evidence to save forests', *Ambio*, no 38, pp125–126

Bren, L. and Hopmans, P. (2007) 'Paired catchments observations on the water yield of mature eucalypt and immature radiata pine plantations in Victoria, Australia', *Journal of Hydrology*, vol 336, pp416–429

Brown, A. E., Podger, G. M., Davidson, A. J., Dowling, T. I. and Zhang, L. (2007) 'Predicting the impact of plantation forestry on water users at local and regional scales: An example for the Murrumbidgee River Basin, Australia', *Forest Ecology Management*, vol 251, pp82–93

Bruijnzeel, L. A. (2004) 'Hydrological functions of tropical forests: Not seeing the soil for the trees?', *Agriculture, Ecosystems and Environment*, vol 104, pp185–228

Buytaert, W., Iniguez, V. and De Bievre, B. (2007) 'The effects of *Pinus patula* forestation on water yield in the Andean paramo', *Forest Ecology Management*, vol 251, pp22–30

Calder, I. R. (2007) 'Forests and water – Ensuring benefits outweigh water costs', *Forest Ecology Management*, no 251, pp110–120

Calder, I. R. and Aylward, B. (2006) 'Forests and floods – In support of an evidence-based approach to watershed and integrated flood management', *Water International*, vol 31, pp544–547

Chandler, D. G. (2006) 'Reversibility of forest conversion impacts on water budgets in tropical karst terrain', *Forest Ecology Management*, vol 224, pp81–94

Chappell, N.A., Tych, W. and Bonell, M., (2007) 'Development of the forSIM model to quantify positive and negative hydrological impacts of tropical reforestation', *Forest Ecology Management*, vol 251, pp52–64

Cornish, P. M. and Vertessy, R. A. (2001) 'Forest age induced changes in evapotranspiration and water yield in a eucalypt forest', *Journal of Hydrology*, vol 242, pp43–63

Cossalter, C. and Pye-Smith, C. (2003) 'Fast-wood forestry: myths and realities', *Forest Perspectives*, CIFOR, Bogor, Indonesia

CSIRO (2008) *Water Availability in the Murray-Darling Basin. A Report to the Australian Government from the CSIRO Murray-Darling Basin Sustainable Yields Project. Water for a Healthy Country Flagship*, CSIRO

De Fraiture, C., Wichelns, D., Roskstrom, J. and Kemp-Benedict, E. (2007) 'Looking ahead to 2050: Scenarios of alternative investment approaches', in D. Molden (ed.) *Water for Food, Water for Life: A Comprehensive Assessment of Water Management in Agriculture*, Earthscan, London

Dudley, N. and Stolton, S. (2003) *Running Pure: The Importance of Forest Protected Areas to Drinking Water*, WWF/World Bank Alliance for Forest Conservation and Sustainable Use, Gland, Switzerland

Dye, P. and Versfeld, D. (2007) 'Managing the hydrological impacts of South African plantation forests: An overview', *Forest Ecology Management*, vol 251, pp121–128

Ellis, T., Hatton, T. and Nuberg, I. (2005) 'An ecological optimality approach for predicting deep drainage from tree belts of alley farms in water-limited environments', *Agricultural Water Management*, vol 75, pp92–116

Ellis, T., Leguédois, S., Hairsine, P. and Tongway, D. (2006) 'Capture of overland flow by a tree belt on a pastured hillslope in south-eastern Australia', *Australian Journal of Soil Research*, vol 44, pp117–125

Enters, T. and Durst, P. (2004) *What Does it Take? The Role of Incentives in Forest Plantation Development in the Asia-Pacific Region*, FAO Regional Office for Asia and the Pacific, Bangkok, Thailand

FAO (2004) 'Global forest resources assessment update 2005: Terms and definitions', Working Paper 83, Food and Agriculture Organization of the United Nations, Rome

FAO (2005a) *Global Forest Resources Assessment*, FAO, Rome

FAO (2005b) 'Forests and floods: drowning in fiction or thriving on facts?', *Forest Perspectives*, FAO, Rome

Farley, K. A., Jobbagy, E. G. and Jackson, R. B. (2005) 'Effects of afforestation on water yield: A global synthesis with implications for policy', *Global Change Biology*, vol 11, pp1565–1576

Fernandez, C., Vega, J. A., Gras, J. M. and Fonturbel, T. (2006) 'Changes in water yield after a sequence of perturbations and forest management practices in an *Eucalyptus globulus* Labill. watershed in Northern Spain', *Forest Ecology Management*, vol 234, pp275–281

Field, C. B., Campbell, J. E. and Lobell, D. B. (2007) 'Biomass energy: The scale of the potential resource', *Trends in Ecology and Evolution*, vol 23, pp65–72

Forrester, D. I. (2007) 'Increasing water use efficiency using mixed species plantations of *Eucalyptus* and *Acacia*', *The Forester*, vol 50, pp20–21

Gallant, J., Van Dijk, A. I. J. M. and Leighton, B. (2005) *FLUSH: A Farm-scale Modelling Framework of Stream Flow and Salinity Changes after Reforestation*, Public CEF Client Report, CSIRO Land and Water

Gerrand, A., Keenan, R. J., Kanowski, P. and Stanton. R. (2003) 'Australian forest plantations: An overview of industry, environmental, and community issues and benefits', *Australian Forestry*, vol 66, pp1–8

Giambelluca, T. W. (2002) 'Hydrology of altered tropical forest', *Hydrological Processes*, vol 16, pp1665–1669

Gutschick, V. P. and BassiriRad, H. (2003) 'Extreme events as shaping physiology, ecology and evolution of plants: Toward a unified definition and evaluation of their consequences', *New Phytologist*, vol 160, pp21–42

Hairsine, P., Croke, J., Mathews, H., Fogarty, P. and Mockler, S. (2002) 'Modelling plumes of overland flow from logging tracks', *Hydrological Processes*, vol 16, pp2311–2327

Hatton, T., Reece, P., Taylor, P. and McEwan, K. (1998) 'Does water efficiency vary among eucalypts in water-limited environments' *Tree Physiology*, vol 18, pp529–536

Hewlett, J. D. (1982) 'Forests and floods in the light of recent investigation', in N. B. Fredericton (ed.) *NB Proceedings, Canadian Hydrology Symposium*, National Research Council of Canada, Ottawa, Canada

Hibbert, A. R. (1969) 'Water yield changes after changing a forest catchment to grass', *Water Resources Research*, vol 5, pp 634–640

Hornbeck, J. W., Adams, M. B., Corbett, E. S., Verry, E. S. and Lynch, J. A. (1993) 'Long-term impacts of forest treatments on water yield: A summary for northeastern USA', *Journal of Hydrology*, vol 150, pp323–344

Ilstedt, U., Malmer, A., Verbeeten, E. and Murdiyarso, D. (2007) 'The effect of afforestation on water infiltration in the tropics: A systematic review and meta-analysis', *Forest Ecology and Management*, vol 251, pp45–51

Jackson, R. B., Jobbágy, E. G., Avissar, R., Roy, S. B., Barrett, D. J., Cook, C. W. and Farley, K. A. (2005) 'Trading water for carbon with biological carbon sequestration', *Science*, vol 310, p1944

Kuczera, G. (1987) 'Prediction of water yield reductions following a bushfire in ash-mixed species eucalypt forest', *Journal of Hydrology*, vol 94, pp215–236

Laurance, W. F. (2007) 'Forests and floods', *Nature*, vol 449, pp409–410

Luce, C. H. (2002) 'Hydrological processes and pathways affected by forest roads: What do we still need to learn?', *Hydrological Processes*, vol 16, pp2901–2904

McJannet, D. L. and Vertessy, R. A. (2001) 'Effects of thinning on wood production, leaf area index, transpiration and canopy interception of a plantation subject to drought', *Tree Physiology*, vol 21, pp1001–1008

McVicar, T. R., Li, L., Van Niel, T. G., Zhang, L., Li, R., Yang, Q., Zhang, X., Mu, X., Wen, Z., Liu, W., Zhao, Y., Liu, Z. and Gao, P. (2007) 'Developing a decision support tool for China's re-vegetation program: Simulating regional impacts of afforestation on average annual streamflow in the Loess Plateau', *Forest Ecology and Management*, vol 251, pp65–81

Mackensen, J., Klinge, R., Ruhiyat, D. and Fölster, H. (2003) 'Assessment of management-dependent nutrient losses in tropical industrial tree plantations', *Ambio*, vol 32, pp106–112

Makarieva, A. M., Gorshkov, V. G. and Li, B. (2009) 'Precipitation on land versus distance from the ocean: Evidence for a forest pump of atmospheric moisture', *Ecological Complexity*, vol 6, pp302–307

Malmer, A. (1996) 'Hydrological effects and nutrient losses of forest plantation establishment on tropical rainforest land in Sabah, Malaysia', *Journal of Hydrology*, vol 174, pp129–148

May, B. P., Smethurst, C., Carlyle, D., Mendham, Bruce, J. and Baillie, C. (2009) *Review of Fertiliser Use in Australian Forestry*, Forest & Wood Products, Australia

Meesters, A. G. C. A., Dolman, A. J., and Bruijnzeel, L. A. (2009) 'Comment on "Biotic pump of atmospheric moisture as driver of the hydrological cycle on land" ' by A. M. Makarieva and V. G. Gorshkov, Hydrol. Earth Syst. Sci., vol 11, pp1013–1033, 2007, Hydrol. Earth Syst. Sci., vol 13, pp1299–1305

Nepstad, D. C., de Carvalho, C. R., Davidson, E. A., Jipp, P. H., Lefebvre, P. A., Negreiros, H. G., da Silva, E. D., Stone, T. A., Trumbore, S. E. and Vieira, S. (1994) 'The role of deep roots in the hydrological and carbon cycles of Amazonian forests and pastures' *Nature*, vol 372, pp666–669

Newman, B. D., Wilcox, B. P., Archer, S. R., Breshears, D. D., Dahm, C. N., Duffy, C. J., McDowell, N. G., Phillips, F. M., Scanlon, B. R. and Vivoni, E. R. (2006) 'Ecohydrology of water-limited environments: A scientific vision', *Water Resources Research*, vol 42

Nykvist, N., Grip, H. and Liang, S. B. (1994) 'Nutrient losses in forest plantations in Sabah, Malaysia', *Ambio*, vol 23, pp210–215

Oki, T. and Kanae, S. (2006) 'Global hydrologic cycle and world water resources', *Science*, vol 313, pp1068–1072

Pannell, D. J. (2008) 'Public benefits, private benefits, and policy mechanism choice for land-use change for environmental benefits', *Land Economics*, vol 84, pp225–240

Peichl, M. and Altaf Arain, M. (2006) 'Above- and belowground ecosystem biomass and carbon pools in an age-sequence of temperate pine plantation forests', *Agricultural and Forest Meteorology*, vol 140, pp51–63

Perrot Maitre, D. and Davis, P. (2001) *Case Studies of Markets and Innovative Financial Mechanisms for Water Services from Forests*, Forest Trends, Washington, DC

Petheram, C., Walker, G., Grayson, R., Thierfelder, T. and Zhang, L. (2002) 'Towards a framework for predicting impacts of land-use on recharge: A review of recharge studies in Australia', *Australian Journal of Soil Research*, vol 40, pp397–417

Pielke Sr., R. A., Beltra'n-Przekurat, A., Hiemstra, C. A., Lin, J., Nobis, T. E., Adegoke, J., Nair, U. S. and Niyogi, D. (2006) *Impacts of Regional Land Use and Land Cover on Rainfall: An Overview*, IAHS Publications, Havana, Cuba,

Pitman A. J., Narisma, G. T., Pielke, R. A., Sr. and Holbrook, N. J. (2004) 'Impact of land cover change on the climate of southwest Western Australia', *Journal of Geophysical Research*, vol 109

Putuhena, W. M. and Cordery, I. (2000) 'Some hydrological effects of changing forest cover from eucalyptus to *Pinus radiata*', *Agricultural and Forest Meteorology*, vol 100, pp59–72

Roberts, J. (1999) 'Plants and water in forests and woodlands', in A. J. Baird and R. L. Wilby (eds) *Ecohydrology*, Routledge, London

Scott, D. F., Bruijnzeel, L. A. and Mackensen, J. (2005) 'The hydrological and soil impacts of forestation in the tropics', in M. Bonell and L. A. Bruijnzeel, (eds) *Forests, Water and People in the Humid Tropics*, Cambridge University Press, Cambridge

Shiklomanov, I. A. (1997) *Assessment of Water Resources and Water Availability in the World*, World Meteorological Organization/Stockholm Environment Institute, Geneva

Shortle, J. S. and Horan, R. D. (2008) 'The economics of water quality trading', *International Review of Environmental and Resource Economics*, vol 2, pp101–133

Sidle, R. C., Ziegler, A. D., Negishi, J. N., Abdul Rahim, N., Siew, R. and Turkelboom, F. (2006) 'Erosion processes in steep terrain – truths, myths, and uncertainties related to forest management in Southeast Asia', *Forest Ecology and Management*, vol 224, pp199–225

Silberstein, R.,Vertessy, R., McJannet, D. and Hatton, T. (2002) 'Tree belts on hillslopes', in R. Stirzaker, R. Vertessy and R. Sarre (eds) *Trees, Water and Salt: An Australian Guide to Using Trees for Healthy Catchments and Productive Farms*, RIRDC, Canberra, Canada

Stednick, J. D. (1996) 'Monitoring the effects of timber harvest on annual water yield', *Journal of Hydrology*, vol 176, pp79–95

Swank, W. T., Swift, L. W., Jr. and Douglas, J. E. (1988) 'Streamflow changes associated with forest cutting, species conversions, and natural disturbances', in W. T. Swank and D. A. Crossley (eds) *Forest Hydrology at Coweeta*, Ecological Studies

Trimble, S. W., Weirich, F. H. and Hoag, B. L. (1987) 'Reforestation and the reduction of water yield on the Southern Piedmont since circa 1940', *Water Resources Research*, vol 23, pp425–437

US EPA (2001) *The United States Experience with Economic Incentives for Protecting the Environment*, Report number EPA-240-R-01-001, Environment Protection Agency, Washington, DC

Valente, C., Manta, A. and Vaz, A. (2004) 'First record of the Australian psyllid *Ctenarytaina spatulata* Taylor (Homoptera: Psyllidae) in Europe', *Journal of Applied Entomology*, vol 128, pp369–370

Vanclay, J. K. (2008) 'Managing water use from forest plantations', *Forest Ecology and Management*, vol 257, pp385–389

Van der Molen, M. K., Dolman, A. J., Waterloo, M. J. and Bruijnzeel, L. A. (2006) 'Climate is affected more by maritime than by continental land use change: A multiple scale analysis', *Global and Planetary Change*, vol 54, 128–149

Van Dijk, A. I. J. M. and Bruijnzeel, L. A. (2001) 'Modelling rainfall interception by vegetation of variable density using an adapted analytical model. 2: Model validation for a tropical upland mixed cropping system', *Journal of Hydrology*, vol 247, pp239–262

Van Dijk, A. I. J. M. and Keenan, R. (2007) 'Planted forests and water in perspective', *Forest Ecology and Management*, vol 251, pp1–9

Van Dijk, A. I. J. M., Hairsine, P. B., Peña Arancibia, J. and Dowling, T. I. (2007) 'Reforestation, water availability and stream salinity: A multi-scale analysis in the Murray-Darling Basin, Australia', *Forest Ecology and Management*, vol 251, pp94–109

Van Dijk, A. I. J. M., Van Noordwijk, M., Bruijnzeel, L. A., Calder, I. R., Schellekens, J. and Chappell, N. A. (2008) 'Forest–flood relation still tenuous – comment on "Global evidence that deforestation amplifies flood risk and severity in the developing world" by C. J. A. Bradshaw, N. S. Sodi, K. S.-H. Peh and B. W. Brook', *Global Change Biology*, vol 15, pp110–115

Vertessy, R. A., Watson, F. G. R. and O'Sullivan, S. K. (2001) 'Factors determining relations between stand age and catchment water yield in mountain ash forests', *Forest Ecology and Management*, vol 143, pp13–26

Vertessy, R. A., Zhang, L. and Dawes, W. R. (2003) 'Plantations, river flows and river salinity', *Australian Forestry*, vol 66, pp55–61

Walker, J. and Reddell, P. (2007) 'Retrogressive succession and restoration in old landscapes', in L. Walker, J. Walker, and R. Hobbs (eds) *Linking Restoration and Ecological Succession*, Springer Science, New York

Waterloo, M. J., Schellekens, J., Bruijnzeel, L. A. and Rawaqa, T. T. (2007) 'Changes in catchment runoff after harvesting and burning of a *Pinus caribaea* plantation in Viti Levu Fiji', *Forest Ecology and Management*, vol 251, pp31–44

Weathers, K. C., Lovett, G. M. and Likens, G. E. (1995) 'Cloud deposition to a spruce forest edge', *Atmospheric Environment*, vol 29, pp665–672

White, D. A., Beadle, C. L., Battaglia, M., Benyon, R. G., Dunin, F. X. and Medhurst, J. L. (2001) 'A physiological basis for management of water use by tree crops', in E. K. S. Nambiar and A. G. Brown (eds) *Plantations, Farm Forestry and Water: Proceedings of a National Workshop*, Agroforestry, Melbourne

Wilson, K. B., Hanson, P. J., Mulholland, P. J., Baldocchi, D. D. and Wullschleger, S. D. (2001) 'A comparison of methods for determining forest evapotranspiration and its components: Sap-flow, soil water budget, eddy covariance and catchment water balance', *Agricultural and Forest Meteorology*, vol 106, pp153–168

Wullschleger, S. D., Meinzer, F. C. and Vertessy, R. A. (1998) 'A review of whole plant water use studies in trees', *Tree Physiology*, vol 18, pp499–512

Zhang, L., Hickel, K., Dawes, W. R., Chiew, F. and Western, A. (2004) 'A rational function approach for estimating mean annual evapotranspiration', *Water Resources Research*, vol 40

Zhang, L., Vertessy, R., Walker, G., Gilfedder, M. and Hairsine, P. (2006) *Afforestation in a Catchment Context – Understanding the Impacts on Water Yield and Salinity*, CSIRO Land and Water Science Report

5

Silvicultural options to enhance and use forest plantation biodiversity

Jürgen Bauhus and Joachim Schmerbeck

Introduction

While the total forest area of the world still declines, the area of tree plantations is growing rapidly (Chapter 1). Although plantations occupy only 4 per cent of the global forest area (FAO, 2005) they dominate the landscape in some parts of the world (Sedjo, 1999). Most forest plantations are primarily established for the efficient production of wood products such as timber, pulp and charcoal, At the same time, plantations increasingly supply other services as well (Cossalter and Pye-Smith, 2003). Plantations can also be established to serve other aims like carbon sequestration, erosion control, water regulation and biodiversity conservation (Chapters 3 and 4; Norton, 1998; Scott et al, 2005; Byrne and Milne, 2006). The provision of these ecosystem services through plantations has also been integrated in community forest programmes that aim to reduce the pressure on natural forests (e.g. Seeland, 1999; Klooster and Masera, 2000). As the area in natural forests declines, ecosystem goods and services supplied by plantations become more and more important as a substitute for services of natural forests. The provision of habitat for plants and animals through plantations or tree cultivation in agroforestry systems is one important service that has received increasing attention (Schroth and da Mota, 2004; Carnus et al, 2006), particularly in areas where native forests have become rare (e.g. Berndt et al, 2008).

The provision of ecosystem goods and services, such as habitat, is linked to ecosystem functioning (Chapter 2). Ecosystem functioning has been described as the 'capacity of natural processes and components to provide goods and services that satisfy human needs, directly or indirectly' (de Groot, 1992). Ecosystem functioning depends to a large extent on the stability and complexity of its structures, composition and processes (Noss, 1990; Naeem et al, 2002). The continuing loss of biodiversity has prompted concerns that the functioning of ecosystems may become impaired and thus compromise the services humans derive from ecosystems (Daily 1997; MEA, 2005b). In the past, there has been much research on the effects of environmental changes or management of natural resources on biodiversity (e.g. Bawa and Seidler, 1998). Now, the emphasis of biodiversity research has shifted towards the functional role of biodiversity, the biodiversity effects on ecosystem functioning (Hillebrand and Matthiesen, 2009). This new ecological framework stresses the active role of the biota and its diversity on the functioning of ecosystems and subsequent effects on the provisioning of ecosystem services (Chapin et al, 2000; Naeem, 2002; Hillebrand and Matthiesen, 2009). In other words, 'the diversity of life determines the manner in which life alters the environment, much as if diversity were a catalyst to life's biogeochemical activities' (Naeem, 2002). With regard to plantation forests, the role of tree species diversity for the functioning of forest ecosystems has been increasingly recognized (Scherer-Lorenzen et al, 2005). Thus in addition to reviewing the role of plantations as harbours of biodiversity in the second part of this chapter, we will first explore the role of biodiversity in plantations for the provisioning of other ecosystem goods and services such as the provision of non-timber forest products, the maintenance of nutrient cycling, carbon sequestration, etc. The importance of biodiversity for ecosystem functioning, for example the increase in productivity observed in many tree species mixtures (e.g. Forrester et al, 2006; Kelty, 2006), may be a stronger incentive for plantation owners to maintain or increase biodiversity than the provisioning of habitat. In addition, understanding the role of diversity in plantations may help to optimize their design to meet ecosystem functions and thus ecosystem services.

Forest plantations are commonly very simple ecosystems in terms of their compositional and structural diversity and are subject to intensive management interventions. Most plantations have been established with the explicit goal of efficient wood production through intensive silviculture, and certain non-timber ecosystem goods and services were basically a by-product of that management. A shift in focus on other non-timber goods and services from plantations is likely to also require changes in the silvicultural approaches used to achieve these broader objectives.

Therefore we ask in this chapter: (1) how is the functioning of plantation ecosystems and the related provisioning of ecosystem goods and services related to diversity within plantations, and (2) how can plantations be silviculturally managed to provide certain levels of biodiversity and to maintain related ecosystem functioning? Following a brief definition of biodiversity and an overview of the biodiversity benefits that may be derived from plantations, we will explore in the first part of the chapter the role of biodiversity, here mostly in terms of tree species diversity, to maintain or enhance the production, regulation and habitat functions of the ecosystem. In the second part of the chapter, we review the literature on the effects of different silvicultural interventions on the diversity of flora and fauna in plantation landscapes and provide recommendations for how plantation management may be changed to accommodate more biodiversity.

Biodiversity and the role of plantations

Today, biodiversity is mostly seen 'as the variability among living organisms from all sources including, inter alia, terrestrial, marine and other aquatic ecosystems and the ecological complexes of which they are part; this includes diversity within species, between species and of ecosystems', or, in short, the 'variety of life on Earth and the natural patterns it forms' (United Nations, 1992). It is widely agreed that biodiversity within and between ecosystems is essential for the provision of many ecosystem goods and services (Daily, 1997; Bawa and Seidler, 1998; Fearnside, 1999; Aretino et al, 2001; Loreau et al, 2002; Scherr et al, 2004; MEA, 2005a; Brown et al, 2006 and many others).

Biodiversity is not just a value in itself; it is of direct and indirect benefit to humanity. The benefits resulting to society from biodiversity can be summarized in the following way (Scherr et al, 2004):

- providing habitat conditions that support diverse wild plant, animal and micro-organism populations of economic, subsistence or cultural value;
- maintaining ecosystem functioning;
- conserving genetic and chemical information of potential future utility;
- providing insurance against future stresses and disturbances;
- providing spiritual, aesthetic and cultural values;
- ensuring the continued existence of wild organisms as legitimate claimants on the Earth's resources.

Most of these biodiversity functions apply to forests (Chapter 2; Carnus et al, 2006). Consequently the value of forest plantations in sustaining and enhancing biodiversity comprises the provisioning of habitat or habitat components, corridor functions and connectivity, buffering of native ecosystems and reducing the pressure on native forests for the extraction of goods.

Biodiversity within plantations will influence the degree of ecosystem functioning and related provision of ecosystem goods and services from plantations. It must be kept in mind, however, that most plantations have been established with the explicit goal of efficient wood production through intensive silviculture. This is gradually changing now, as the role of production ecosystems for the maintenance of sustainable landscapes and the provision of ecosystem services has been brought to public attention (MEA, 2005b). In addition, production systems requiring high external inputs may have to adapt to become more energy efficient and thus rely more on natural processes (e.g. Kirschenmann, 2007).

Moore and Allen (1999) have therefore posed the question of what biodiversity in these plantations actually comprises and what level of diversity managers should aim to maintain or restore. They argue that it is unreasonable to manage for 'biodiversity', when it is not clear to what extent stand and landscape manipulations may benefit some species or hinder others. This argument may be extended to the desired benefits through provision of ecosystem goods and services. Moore and Allen (1999) suggest that the goal of plantation management should be to ensure that the biodiversity, which enhances or does not substantially reduce plantation productivity, be maintained at the stand level and undesired off-site effects be minimized. The latter aspect points to the landscape perspective in which the question of plantation management must be viewed.

The role of plantation diversity on ecosystem functioning

The ecosystem functions and services of forest plantations are practically the same as those obtained from native forests. However, owing to differences in structure and composition, they may be provided to different degrees (Chapters 2, 3 and 4). Plantation forests are largely even-aged stands and thus are characterized by a simple structure. They contain trees of similar age, size and conditions. Tree density is managed to maximize profits, which results in fully or optimally stocked stands in which the crop trees dominate the site. Spatial variability is commonly kept at a minimum to facilitate efficient planning and management, especially in areas where management relies on the use of large machinery. In addition, forest harvesting or other silvicultural interventions do not produce the type of stand structures characteristic of natural ecosystem dynamics (Bauhus et al, 2009). Under these conditions, most variation in stand structure results from differences in stand density and species composition. The effects of stand density are discussed further below under silvicultural interventions, which are the main determinant of stand density. Here, we will focus on the effects of compositional (species) diversity on ecosystem functions and services, including the production, regulating and habitat functions (see Chapter 2).

Today there seems to be a consensus that biodiversity is an important determinant of ecosystem functioning and related ecosystem services (Loreau et al, 2002; MEA, 2005a; Scherer-Lorenzen et al, 2005). This insight is the result of many experiments, observational studies, theoretical models and recent meta-analyses (e.g. Balvanera et al, 2006; Piotto, 2008). However, the generality of the conclusion is based largely on studies of communities of short-lived or fast-growing organisms rather than on studies of long-lived forests. Among other insights from this research, two findings about biodiversity and ecosystem functioning emerged, which are central to the discussion of ecosystem services from plantation forests (Loreau et al, 2002; Hooper et al, 2005; Baumgärtner, 2007):

1 In many situations, ecosystem productivity increases with the level of biodiversity.

2   Biodiversity may enhance ecosystem stability. An increase in the level of biodiversity decreases the temporal variability of the level at which ecosystem services are provided under changing environmental conditions.

The magnitude of both of the above effects declines with the level of biodiversity, so that these functions are tapering off. These two important questions will be dealt with below, when we review the evidence for the relationships between tree species diversity and plantation productivity and resiliency.

The diversity–ecosystem functioning relationships comprise a number of mechanisms, of which the most important with regard to plantation forests are niche complementarity and ecological insurance against stress and disturbances. *Niche complementarity* leads to greater efficiency of local resource exploitation through more diverse communities, in which a mixture of different growth forms of plants can fill the available limited space to a greater extent and thus is able to exploit the available resources better than a monoculture under the same site conditions (Loreau and Hector, 2001; Naeem, 2002). Complementarity exists when different species use different resources, or the same resources but at different times or different points in space, for example different soil layers. In this case, interspecific competition in the mixture is less than intraspecific competition in the monocultures. The concept of niche complementarity also comprises facilitation, the positive interactions among species, where certain species improve environmental conditions for others or supply a critical resource such as nitrogen through fixation. Both of these mechanisms, complementarity and facilitation, often lead to an 'overyielding effect', in which biomass production in species mixtures exceeds the productivity expected on the basis of the yields of the contributing species when grown in monoculture (Ewel, 1986; Loreau et al, 2002; Naeem, 2002; Vandermeer, 1989). However, this applies only when mixed plant communities are growing under the same environmental/site conditions as the monocultures used for comparison. In natural plant communities growing across a range of site conditions, reverse relationships between diversity and productivity may be observed (Naeem, 2002).

The *ecological insurance* concept postulates that more diverse communities are more likely to cope with new conditions when subject to unpredictable stress or disturbance (Yachi and Loreau, 1999). With an increasing number of functionally different species, the probability increases that some of these species can respond in a differentiated manner to the external perturbations or changing environmental conditions. In addition, the probability increases that one species can take over the role of another, redundant species that does not survive the disturbance or new conditions (Walker et al, 1999; Yachi and Loreau, 1999). Following this logic, processes that are carried out by a small number of species are likely to be more sensitive to changes in diversity than those that are carried out by a large number of species.

However, the exact response of ecosystem services to changes in biodiversity through addition to or removal of a species from the ecosystem is not just determined by species richness itself but at least as much by the functional traits of these species (e.g. Tilman et al, 1997). In addition, the patterns of response to different levels of species richness vary for different ecosystem processes and services and between different ecosystems. In most situations, only productivity – which is, however, related to many other ecosystem services – has been investigated.

For terrestrial ecosystems, most of this knowledge has been created through experimentation in fast-growing model systems such as grasslands (e.g. Tilman et al, 1997; Spehn et al, 2005). While there may be a wide empirical knowledge base for the management of mixed-species forests, there is relatively little quantitative information on the effects of diversity on forest ecosystem functioning (Scherer-Lorenzen et al, 2005). In addition, how the effects of diversity may be used to enhance ecosystem functioning in plantation forests has not been reviewed. In the following, we will therefore discuss how the diversity of tree species, mostly expressed as species richness, affects the provision of a range of ecosystem functions. The focus is on tree species richness, since this is the aspect of diversity which is most strongly influenced through management, and because trees are the shaping elements of forest ecosystems and therefore also influence diversity at other scales and trophic levels. We will discuss in particular the influence of tree species diversity on the production, regulation and habitat functions of forest plantations.

Tree diversity effects on ecosystem productivity

The production functions (or provisioning services) comprise the processes that combine and change organic and inorganic substances through primary and secondary production into goods that can be directly used by mankind (Chapter 2; de Groot et al, 2002). In the context of forest plantations, these are primarily the production of woody biomass, commonly of defined qualities, and the production of non-timber goods such as herbs, mushrooms, huntable animals, etc. Whereas the availability of information considering the effects of tree species diversity on production of woody biomass is reasonable (but see Scherer-Lorenzen et al, 2005), the information about non-wood products is scant. Here, we will initially focus on the effect of tree diversity on the production function, because it relates to many of the regulating functions such as carbon, nutrient and water cycling, and because the productive capacity of plantations is the principle motivation for their establishment in most situations (see Chapter 1). Where plantations are not established primarily for the production of woody biomass but for other purposes such as the conservation or restoration of soils, the achievement of these aims will also be influenced by the basic ecosystem functions listed above.

The information about diversity effects on forest ecosystem productivity comes mostly from the analyses of large-scale permanent forest inventory plots representing gradients in tree species richness (Caspersen and Pacala, 2001) and from controlled experiments or forest yield plots comparing monocultures and tree species mixtures, mostly, however, two-species mixtures (e.g. Pretzsch, 2005; Forrester et al, 2006; Piotto, 2008). Whereas the first source of data stems mostly from all forms of forests (native, semi-natural and plantations), the latter stems mostly from plantations or semi-natural forests.

So far, there have been only a few studies utilizing an inventory-based analysis of the relationships between tree species diversity and productivity. In two of these studies (Caspersen and Pacala, 2001; Vilà et al, 2007) positive relationships between diversity and productivity have been observed for North American forests and in typical early successional Mediterranean-type forests of Catalonia, Spain, respectively. In the study by Caspersen and Pacala (2001) it was not possible to untangle the confounding effects of site quality on productivity and diversity. Here, cause and effect may actually be reversed, namely that more productive stands simply permit the coexistence of more species. However, Vilà et al (2007) could demonstrate that the positive relationship between tree species diversity and forest productivity was maintained when considering the effects of forest structure, environmental variables and management practices.

In another study by Vilà et al (2003), no effect of species richness was observed in *Pinus sylvestris*-dominated forests, but a positive effect was detected in *P. halepensis* stands of Catalonia. In the latter case, however, tree species richness was no longer a significant factor, when the climate, bedrock types, radiation and successional stage of inventory plots were included in the analysis. The latter study points to the problem of these analyses: the large number of co-variables that may influence the relationship between diversity and productivity (or other function), and the difficulty of accounting for these in a statistically sound way. In addition, forest inventory plots are usually not selected to represent a diversity gradient, and thus most inventory plots are located at the lower end of tree species diversity (Vilà et al, 2007). Furthermore, these types of analyses are not suited to explaining the underlying mechanisms of any diversity–ecosystem function relationship.

In terms of a second source of information about diversity effects on forest ecosystem productivity, most experiments with mixed-species stands also cover only the low end of tree species diversity. They are, however, in contrast to inventory-based analysis, well suited to exploring the underlying mechanisms. Positive effects of tree diversity on productivity occur through competitive reduction, also described as complementarity above, and facilitation (Vandermeer, 1989; Kelty and Cameron, 1995). Productivity in mixtures will out-yield monocultures, when the positive interactions of facilitation and competitive reduction dominate the competitive interactions. Bauhus et al (2006) and Piotto (2008) have shown that in most reported cases, where a statistically valid comparison between mixed- and single-species stands could be made, the productivity was equal or greater in mixed stands, in particular, if nitrogen-fixing species were admixed (see also Forrester et al, 2006) (Figure 5.1). However, it has to be kept in mind that cases where mixtures are inferior

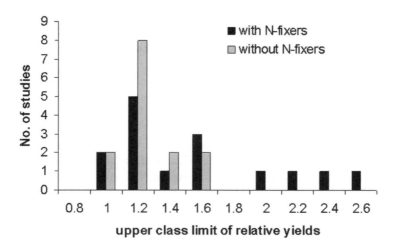

Figure 5.1 *Frequency of relative yield totals of mixed-species plantations reported in the literature*

Note: The relative yield total (RYT), provides the factor by which the productivity in a mixture is superior or inferior when compared to a monoculture; e.g. an RYT of 1.4 indicates that 1.4ha of monocultures (0.7ha of each species) would have to be planted to obtain the same yield as in 1ha of a 50:50 mix of the same species.
Source: after Bauhus et al, 2006

in productivity to monocultures of the respective companion species, are probably less likely to be reported in the literature.

The most important mechanisms that lead to increased productivity in tree mixtures comprised the following processes and patterns.

Canopy stratification

Successful (meaning more productive than respective monocultures) mixed-species plantations often have stratified canopies with a fast-growing shade-intolerant species forming the upper canopy and a more shade-tolerant species forming the lower canopy (Assmann, 1970; Bauhus et al, 2004; Pretzsch, 2005). Shade-intolerant species are capable of higher maximum rates of photosynthesis than more shade-tolerant species. Therefore more efficient use will be made of the higher light intensities in the upper canopy of a mixture than in the upper canopy of a more shade-tolerant species in monoculture. In contrast, shade-tolerant species are capable of maintaining foliage and assimilating at higher rates than shade-intolerant species at lower light intensities (Kelty, 1992). When grown in mixed stands with shade-intolerant species, shade-tolerant species increase the amount of light intercepted compared to monocultures of the shade-intolerant species (e.g. Binkley, 1992; Bauhus et al, 2004). When shade-intolerant species do not form the main canopy, productivity in mixtures may not be higher in mixtures than in monocultures (Parrotta, 1999; Redondo-Brenes and Montagnini, 2006). Canopy stratification alone is not the only factor responsible for the success or failure of these mixed stands, but it is a key factor to ensure the long-term coexistence of species (Forrester et al, 2006). Stratified canopies also offer different habitat niches with consequent effects on dependent faunistic diversity (see below).

Below ground stratification
Competition may be reduced also below ground through physical or chemical stratification of roots. Physical stratification comprises differences in fine-root distribution that affect exploitation strategies. This can be in the form of the layering of root systems (e.g. Schmid and Kazda, 2002) or different fine-root architectures that can be associated with different soil exploration and exploitation strategies (e.g. Bauhus and Messier, 1999). Chemical stratification may occur when co-occurring species take up different forms of nutrients, for example ammonium vs. nitrate or organic N, or employ different acquisition strategies for nutrient uptake through different types of mycorrhiza (Ewel, 1986; Schulze et al, 1994).

Facilitation
Facilitation among tree species may occur through direct amelioration of harsh environmental conditions or increasing the resource availability, or indirectly through the introduction of beneficial organisms (mycorrhizae and other soil microbes) or protection from herbivores (Callaway, 1995; Forrester et al, 2006). Increased resource availability in mixed-species stands may result from accelerated cycling of nutrients, for example in the forest floor, the mineral soil or in biomass. This has been observed for example for N in mixed stands of Sitka spruce (*Picea sitchensis*) and Scots pine (*Pinus sylvestris*) in Scotland (Millar et al, 1986 in Williams, 1996) or P in mixtures of *Eucalyptus globulus* with *Acacia mearnsii* (Forrester et al, 2005). The increase in the pool of available nutrients or in the rate of its cycling may result from stratification of the root systems, or the accelerated decomposition of organic material in a mixture, which has been observed in the majority of cases, when litter of different species was mixed (Gartner and Cardon, 2004; Hättenschwiler, 2005). As indicated in Figure 5.1, fixation of atmospheric N by one of the co-occurring species often has a large impact on N availability and thus productivity. The greatest increases in productivity in mixed stands when compared to mono-specific stands have been found in mixtures with N-fixing species (see also Kelty and Cameron, 1995; Forrester et al, 2006). Once fixed through symbiotic bacteria or actinomycetes, N can be transferred from N_2-fixing species to the non-N_2-fixing species via litter decomposition and subsequent mineralization of organic N.

The relationship between tree species diversity and productivity as well as between diversity and some of the other ecosystem processes such as decomposition and nutrient cycling is, however, idiosyncratic. The performance of the mixtures in comparison to respective monocultures is often more strongly influenced by the specific species combinations and site conditions than by the diversity (Forrester et al, 2005; Pretzsch, 2005; Redondo-Brenes and Montagnini, 2006). It is therefore difficult to forecast the outcome of the various interactions in mixed stands, also because these interactions may change as the stands develop (e.g. Forrester et al, 2004). In relation to site conditions, the success of mixtures seems to depend on whether nutrient or water limitations of the companion species can be reduced in the mixture (e.g. Laclau et al, 2008; Forrester et al, 2010). However, there is some indication that mixtures of species representing complementary functional types (e.g. shade-tolerant with shade-intolerant, early- and late-successional, or N_2-fixing and non-fixing species) result in most cases in higher productivity when compared to monocultures (Kelty, 1992). The uncertainties regarding the expected outcome in terms of productivity of mixed-species plantations are certainly one of the reasons for the low uptake of mixed plantations in practice. This suggests that to reduce this uncertainty, the relationships between functional traits of tree species in mixtures in relation to ecosystem functioning should be the focus of future research.

In most types of plantations, however, the goal is not only to produce a certain amount of woody biomass but also to produce a certain quality of wood for industrial end-uses. This is a very important aspect of the production function of species-diverse plantations, since most forest owners will only adopt higher levels of fine-scaled tree species diversity, if mixtures produce at least the same quality of timber as mono-specific stands. Properties such as wood density, lignin content and fibre length are unlikely to be influenced by tree species mixtures. The quality of trees for solid wood products depends, in addition to year-ring structure and internal rot, most importantly on the criteria dimension, taper, straightness and proportion of branch-free wood (e.g. Cutter et al, 2004). These determinants of tree quality are mainly driven by crown dynamics, which may be strongly influenced by species mixtures. For example, branchiness, which is the most important criterion for downgrading of timber (Montagu et al, 2003), is closely related to crown size, which is commonly closely related to the competition experienced by subject trees (e.g. Alcorn et al, 2007). Thus tree quality parameters of subject trees are dependent on interactions with neighbouring trees, which may be highly influenced by the tree species composition and diversity of the neighbourhood (Sumida et al, 2002). However, very little is known about this, since most studies examining timber quality have focused on mono-specific stands (e.g. Alcorn et al, 2007; Hein et al, 2008).

Previous studies on tree species diversity have focused on stand-level productivity (Pretzsch, 2005; Forrester et al, 2007), but not on the quality of trees grown in mixtures. However, the interactions observed between trees in these studies already point to possible crown interactions. Hetero-specific trees may pose less, equal or more competition than neighbouring con-specific trees. This will depend on species traits influencing competition such as height growth rates, crown expansion rates, shade tolerance, stiffness of branches, protection of buds against abrasion, timing of leaf flush, etc. (e.g. Sumida et al, 2002). Differences in growth rates, where subject trees grow slower than hetero-specific neighbours, may affect the dimension and straightness of stems, in particular in phototrophic species with low apical dominance. As a result of spatially heterogeneous competition, crowns may become one-sided (or imbalanced) (Jones and Harper, 1987) leading to tension and compression wood (Bowyer at al, 2003). In situations where neighbouring species are weak competitors, the green crown of subject trees may rise more slowly leading to larger, lower branches and delayed self-pruning. In contrast, improved self-pruning and reduced development of epicormic branches may result where neighbours are shade-tolerant species of lower height. Often mixed-species stands result from the unplanned colonization of monocultures through voluntary regeneration of native species. These spontaneous mixtures, which may only be temporary, can have beneficial effects on form and quality, where they increase stand density without outcompeting the crop tree species (Valkonen and Ruuska, 2003). This may not be an effect of diversity but of stand density. However, spontaneous mixtures, in particular with native species, are likely to also have beneficial effects on biodiversity (see below).

Given the interest in mixed-species forests in many regions, as can be seen in efforts to convert coniferous monocultures into mixed deciduous–coniferous forests in Europe (Spiecker et al, 2004), this knowledge gap on the influence of tree species diversity on timber quality is surprising. The available information suggests that the quality of trees may not have to be compromised in mixed stands, if the species composition is judiciously chosen and managed. This points also to the need for information about the traits of companion tree species to make the outcomes of mixed-species plantations more predictable.

Provisioning of non-wood forest products in diverse plantations

In addition to the production of wood, the production function of plantations also extends to other plant products and animals that may be harvested for human consumption. Forest products that are not related to timber have been important to human beings since the hunting and gathering age. The importance of non-wood forest products (NWFPs) for the livelihood systems of people all over the world, in developing as well as industrialized countries, has been clearly pointed out in the literature (Wickens, 1991; SCBD, 2001; Chamberlain et al, 2002; Ticktin, 2004; Kaushal and Melkani, 2005; Emery et al, 2006). Here, the term non-wood forest products (NWFPs) will be used for all the biological material (other than wood products) that can be utilized within the household, be marketed or have social, cultural or religious significance (Wickens, 1994). Typically this includes nuts, fruits, berries, mushrooms, herbs, bark, resin, rubber, etc. This definition does not include forage, which sustains livestock or game animals (Wickens, 1994). However, in the livelihood systems of many people, forage from forested land plays an important role.

With a shrinking area of natural forests, the supply of NWFPs is shrinking accordingly. For example, it has been estimated that the demand for these products outstrips the sustainable yield from Indian forests (Appasamy, 1993), and in countries such as Nepal, some non-timber forest product species are threatened by extinction from overexploitation (Maraseni, 2008). Not surprisingly, NWFPs are cultivated at an industrial scale in plantations, for example for rubber, palm oil, nuts, etc. (Last, 2001). In addition, the active management of forests for provision of NWFPs has commenced in some areas, partly to reduce the pressure on native forest (e.g. Maraseni, 2008; Trauernicht and Ticktin, 2005). This objective, which is similar to that for the establishment of tree plantations in some regions, suggests that there may be some synergies. However, the active cultivation of NWFPs is certainly more in the domain of agroforestry, in particular in home gardens, which provide a rich variety of these products and function as harbours of biodiversity (Kumar and Nair, 2006). The relationship between biodiversity and NWFPs in these systems warrants a separate discussion, which is beyond the scope of this chapter and this book. Here, we ask, whether management for diversity in 'conventional' tree plantations might accommodate the provisioning of NWFPs. However, studies that deal specifically with silvicultural approaches to maintaining or enhancing the provision of NWFPs are extremely rare. This lack of knowledge has been identified by different authors in different regions. Neumann and Hirsch (2000) reviewed the literature about the policies and practices regarding the management of NWFPs and found, that 'there are promising speculations about integrating timber and NWFP management, but few studies that experimentally test the possibilities of different management techniques and silvicultural treatments, and there is even less information specific to plantations. Chamberlain et al (2002) analysed the management plans of 32 eastern national forests of the US concerning the space given for the description and management of NWFPs. Despite an increasing importance of various NWFPs in the studied states, NWFPs were mentioned in only seven plans and no plan devoted more than 1 per cent of its text to them. The lack of explicit concern for NWFPs in management plans, inventory, policy and legislation is echoed by others (Lynch and McLain, 2004; Emery et al, 2006). The lack of such information is particularly obvious for tropical countries, where NWFPs play a more important role for local livelihoods (Panayotou and Ashton, 1992; Mahapatra and Mitchell, 1997; Gautam, 2001; Ticktin et al, 2002; Ticktin, 2004).

One of the speculations about silvicultural treatments and NWFPs mentioned by Neumann and Hirsch (2000) is that product yield in forests or plantations may be increased through the enhancement of biodiversity and the creation of structural diversity (Gautam, 2001; Ticktin, 2004; Gautam and Devoe, 2006). However, this often seems to be largely related to the interactions between canopy light regime and understorey development. For example, Gautam (2001) found that lopping and litter removal in sal (*Shorea robusta*) forests in Nepal increased understorey light conditions and thus enhanced the supply of NWFPs such as herbs, lianas and grasses, whereas useful shrubs declined. Floristic diversity in plantations, which is probably related to the diversity of NWFPs, is often inversely related to canopy cover (e.g. Bone et al, 1997; and discussion below). However, based on existing studies, the effects of tree species diversity on understorey diversity can usually not be disentangled from the effects of succession and plantation age (see Box 5.2). For example, monoculture plantations have been proven to be a probate system for the successful re-establishing or restoration of diverse native vegetation (Lugo, 1997; Parrotta et al, 1997). Here, functional traits such as shade-tolerance, litter decomposability or the allelopathic effects of the tree species are more important than questions of tree species richness or origin, native vs. exotic.

Most studies in the literature are concerned with the management of selected non-timber products. The question that is important for this review, namely to what extent production of NWFPs can be accommodated in plantations grown for other purposes without affecting the main production goals or what the trade-offs may be between the provision of NWFPs and wood production, is not dealt with. Despite the lack of information, however, some general statements can be made. The provision of mushrooms would be highly compatible with plantation management, particularly when these are the fruiting bodies of mycorrhizal fungi, which are fed by the tree crop. However, these may not necessarily be native mushrooms, since for some exotic species the mycorrhizae were imported as well (Brundrett et al, 1995). For the provision of NWFPs that stem from understory vegetation (fruits, berries, medicinal plants, roots and tubers) the same considerations apply, discussed further below in relation to the richness and diversity of understorey vegetation in plantations. The provision of these plant products may depend on the functional traits of tree species, their canopy light transmission, competition for soil resources, litterfall, initial site preparation, etc. Thus provision NWFPs may be influenced by spacing, thinning and the initial establishment techniques.

Information about the role of plantations in the provisioning of huntable animals is even scarcer, although large-bodied mammals and birds are a major source of dietary protein, in particular in many parts of the tropics (Robinson and Bennett, 2004). This lack of information is probably related to the scale of investigation, since the home range of many of these animals is defined by landscapes comprising a variety of land use and forest types, and where it is difficult to relate hunting and habitat preferences to particular landscape units such as plantations. The habitat quality for huntable animals is likely to depend on the structural diversity of plantation forests as well as the diversity of the landscape and the proportion of native ecosystems in the landscape (e.g. Nasi et al, 2008). One study, which investigated subsistence hunting patterns in a landscape with different land-use types in tropical Brazil, showed that most of the kills were sourced from primary forest (Parry et al, 2009). In contrast, hunting pressure in plantations was low, despite a high catch-per-unit-effort. The authors assume that plantations were less attractive to hunters because there were limited additional benefits from visiting these habitats. Because large-scale tree plantations fail to attract hunters away from primary forests and support few forest specialist animals, they were viewed to have limited conservation potential for large vertebrates.

Tree diversity effects on regulating functions

In addition to the provisioning functions for wood and non-wood forest products, plantations have important regulating functions for water, nutrient and carbon cycles as well as for ecosystem resiliency. Although the role of plantations for hydrological cycles and carbon storage and sequestration have been highlighted previously in Chapters 3 and 4, we will here return to the question of how biodiversity, with a focus on tree species richness, might influence the ecosystem functioning related to these services.

Water

Despite the fact that water is a fundamental and critical resource economically, environmentally and socially (Calder, 2005), the effect of tree species mixtures on forest hydrological cycles has been largely ignored. This is surprising, since as we have seen above, mixed-species forests are often more productive than their mono-specific counterparts and forest water use does normally increase with productivity (Law et al, 2002). Therefore it should be asked whether mixtures use significantly more water than monocultures on the same site. However, there are only a few studies that have compared the water use of trees in mixtures to monocultures (Schume et al, 2004; Anders et al, 2006; Forrester et al, 2010). Forrester et al (2010) found that water use in mixtures was higher than in monocultures, although the water-use efficiency of the companion species had also increased. Similarly, water use in mixed stands of Norway spruce and European beech was substantially higher than in pure Norway spruce stands (Schume et al, 2004). However, in mixtures and monocultures of *Pinus sylvestris* and *Fagus sylvatica*, groundwater recharge was higher under mixtures (Anders et al, 2006) than under pine, despite a higher water use of trees in the mixture. The improved soil moisture and drainage in the mixed stand was the result of suppression by the shade-casting beech of a dwarf-shrub and grassy understorey with high transpiration. This study shows that interactions in mixed stands can be complex and go beyond the direct influence of one tree species on the other. However, these findings demonstrate that there may be trade-offs between the use of tree species diversity for other functions and the provision of water or the susceptibility to drought stress.

Carbon sequestration and storage

A functional relationship between tree species diversity and C storage and sequestration in plantation forests would have important implications for the management of the C-sinks in reforestation and afforestation projects (UNFCCC, 2005). The effects of tree species diversity on above ground sequestration of C in plantations is closely related to that of ecosystem productivity, which has been discussed above (see also Chapter 3). In addition, tree species diversity can have effects on below ground and soil carbon through effects on litter quality and decomposition (Giardina et al, 2001; Binkley and Menyailo, 2005) and also the diverse rooting patterns of trees leading to the deposition of organic material at various soil depths (Johnson, 1992). There are some indications of the positive effects of plant species richness on soil organic carbon storage in forest or agroforest ecosystems (Chen, 2006; Saha et al, 2009). These observational studies cannot clearly separate the effects of species diversity from that of all the other environmental influences. However, Gleixner et al (2005) suggested that these are likely to be indirect effects of plant diversity mediated by feedbacks between above ground and below ground diversity, and effects on water cycles and decomposition processes. Thus species-specific influences on these processes or those that are related to functional types are likely to have a large influence on soil C storage. One example for such an effect is the inclusion of N-fixing species leading to increased soil organic carbon sequestration, presumably through reduced mineralization of N-enriched soil organic C under N-fixing species when compared to non-N-fixing species (Resh et al, 2002).

New insights in the effects of biodiversity on soil C sequestration and nutrient cycling can be expected from controlled manipulative experiments of tree species diversity (e.g. Scherer-Lorenzen et al, 2007). In addition, an indirect effect of tree diversity on forest C storage may be through increased resistance against disturbances (see also Chapter 3).

Nutrient cycling

Mixing tree species may lead to increased availability and more efficient use of nutrients through accelerated cycling or through increased capture of nutrients. The question of nutrient cycling is of paramount importance in high-yielding plantations, where the export of nutrients with harvested products may be very high (Gonçalves et al, 1997). Any processes and mechanisms that lead to reduced nutrient losses per unit of exported biomass and to reduced fertilizer inputs, in particular of nutrients with limited supply such as P, will help to improve the sustainability of plantations. How changes in tree diversity, from monocultures to mixtures of varying species richness and composition, affect nutrient cycling is still rather equivocal (Scherer-Lorenzen et al, 2005). Interspecific differences in resource capture can be attributed to different resource requirements, uptake abilities and niches occupied (Rothe and Binkley, 2001). Hence, mixing species with such different traits and resource niches may lead to niche differentiation and resource partitioning, resulting in increased resource use complementarity. A recent meta-analysis showed that in the majority of cases the above ground nutrient content and nutrient use efficiency (N and P) of species grown in mixtures were higher than in monocultures, indicating an increase in the proportion of resources captured from a site (Richards et al, 2010). However, this study also indicated that in a substantial number of cases there is either no effect or a negative effect. This again points to the need for careful selection of companion species to obtain the desired influence of mixtures of ecosystem functioning, here the maintenance of site fertility.

Resiliency

Resiliency, as a regulating function of ecosystems, is related to the concept of ecological insurance, which hypothesizes that more diverse communities are more likely to cope with stress or disturbance (Yachi and Loreau, 1999). Tree species diversity may influence both the resistance as well as the resilience in relation to specific disturbance agents, which may be of biotic as well as abiotic nature. Resistance can be defined as the ability of the system to withstand changes of the current state, whereas resilience relates to the capacity of the system to recover from disturbance and to regain the pre-disturbance condition (Attiwill, 1994). For forest ecosystems, the living biomass stock is often used to define this pre-disturbance reference condition.

The most important abiotic disturbances affecting plantations are wind storms and fire. While there are many publications that have analysed the storm resistance of individual tree species and certain stand structures (e.g. Foster, 1988; Schütz et al, 2006), few have considered the question of mixed-species stands (e.g. Lüpke and Spellmann, 1997; Dhôte, 2005), and none have analysed the question of diversity. The resistance to storm is rather species specific and determined by traits such as tree height, rooting patterns, deciduousness vs. evergreen, foliage density, etc. For example, Lüpke and Spellmann (1997) found that Norway spruce when mixed with European beech is no less susceptible to storm damage than in mono-specific spruce stands. However, the mixed stands are more stable, since the beech component is less affected. This implies that these mixtures are not more stable than mono-specific beech stands. Unless the traits conferring wind firmness of the different tree species participating in mixed stands are not influenced by the stand composition, similar results may be expected for other species combinations. The persistence of at least a partial cover of trees following catastrophic storms may, however, be an important advantage for the recovery of stands (Dhôte, 2005).

The issue is much more complicated for fire disturbance, since there are interactions between species traits and disturbance frequency and intensity. Important tree species traits related to fire resistance are: bark thickness, shedding of dead branches, etc. (e.g. Fernandes et al, 2008). In addition, there are species traits promoting fire such as the flammability of leaves and needles as well as litter production (Facelli and Pickett, 1991). Some widely distributed plantation genera such as *Pinus* and *Eucalyptus* are known for their fire promoting traits (Scarff and Westoby, 2006; Ormeno et al, 2009). Thus tree species with low flammability of litter have been commonly used as green fire breaks in plantations (Johnson, 1975). Mixing such species into plantations to accelerate the decomposition of flammable litter, to change the fuel bed properties or to shade a flammable understorey such as grasses may be additional options to reduce fire risks. However, these benefits regarding increased resistance to fire are the result of specific combinations of species, not diversity per se.

In relation to biotic disturbances through pest or pathogen species, there is evidence that lower tree species diversity is related to greater pest insect abundance, density or damage (Jactel et al, 2005). The mechanisms leading to a greater resistance of mixed-species stands comprise reduced accessibility of host trees to pests, a greater impact by natural enemies and the diversion of pests from less susceptible to more susceptible tree species (Jactel et al, 2005). Further, Jactel and Brockerhoff (2007) and Koricheva et al (2006) found that the composition of tree mixtures was likely to be more important than species richness per se. This is because the diversity effects on herbivory were greater when mixed forests comprised taxonomically more distant tree species, and when the proportion of non-host trees was greater than that of host trees. In addition, the effect of tree species diversity is more pronounced for specialist herbivores than for generalists (Koricheva et al, 2006; Jactel and Brockerhoff, 2007). However, there are many situations where the direct comparison of monocultures with polycultures or plantations with native forests has failed to demonstrate that mono-specific stands are more susceptible to diseases (Nair, 2001).

The habitat functions of ecosystems relate to the importance of ecosystems to provide habitat for various stages in the life cycles of wild plants and animals, which, in turn, maintain biological and genetic diversity and evolutionary processes (Chapter 2). Since more tree species can support more dependent species such as micro-organisms and invertebrates, the benefits for biodiversity are obvious, in particular, if native species are components of mixed-species plantations (Hartley, 2002). This other potentially important regulating function of plantations is dealt with below, when we discuss the silvicultural options to maintain or enhance biodiversity in plantation forests, since the effects of plantation management on biodiversity are of particular public concern (e.g. Brown et al, 2006).

Silvicultural options to increase plantation biodiversity

In the previous section we have seen that biodiversity at the level of tree species richness has the potential to enhance plantation ecosystem functioning. As a result the related ecosystem goods and services may be provided at a higher level when compared to mono-specific plantations. While the relationships between tree species diversity and the provisioning and regulating functions may not in every case be positive, there are only a few examples where these may be detrimental. The examples above indicate that there may be economic incentives for plantation owners to incorporate higher levels of biodiversity to improve productivity, to sequester more C or to reduce the susceptibility to pests, pathogens or fire. However, at other trophic levels, and with regard to herbivores or predators, the maintenance of biodiversity may not be of direct benefit to plantation owners and may even come at a cost, if the provision of habitat for these organisms requires additional or more intensive silvicultural operations or reduces yield. In this section, we will review the current knowledge about silvicultural options to maintain or enhance plantation biodiversity.

Silviculture is the manipulation of forest structure and dynamics at the stand level with the specific aim of producing certain goods or services. While silvicultural systems operate at the stand level, they have ramifications for the higher levels of forestry planning and management (Nyland, 2002). Likewise, silvicultural decisions at the stand level are influenced by or have to conform with management goals at the higher levels of management such as the estate or the landscape. For example, the establishment of a potentially invasive exotic tree species in some plantation stands can have effects on biodiversity at the landscape scale, or the management of short rotations in plantation stands may provide the highest financial return, but may impact negatively on catchment water run-off (Nyland, 2002).

There is a range of silvicultural options that forest managers can apply at the stand level to achieve different management goals. They can be grouped into the following five categories:

1   site preparation and residues management;
2   tree species selection and mixtures;
3   stand density management (spacing and thinning);
4   structural complexity creation;
5   rotation length.

Silvicultural options to establish, maintain or enhance biodiversity values exist at the different planning levels (see Kerr, 1999, for a discussion of the options). Habitat or habitat components can be influenced at the tree neighbourhood level (Coates and Burton, 1997), stand level (e.g. Geldenhuys, 1997; Hartley, 2002; Lindenmayer and Franklin, 2002; Watt et al, 2002; Montes et al, 2005; Brown et al, 2006) as well as the landscape level (e.g. Lindenmayer and Hobbs, 2004; Tubelis et al, 2004; Montes et al, 2005), while the matrix and corridor function of plantations and the buffering of native ecosystems is clearly an issue for estate or landscape planning (Bone et al, 1997; Tubelis et al, 2004; Nasi et al, 2008). Therefore, silvicultural prescriptions are just one element in an overall plan to achieve certain biodiversity goals in plantations (Figure 5.2).

Plantations may harbour a surprisingly high proportion of native plant biodiversity. For example, Keenan et al (1997) found over 300 plant species beneath tropical timber plantations of the exotic *Pinus caribaea* and the natives *Araucaria cunninghamii*, *Flindersia brayleyana* and *Toona ciliata* in Northern Queensland. In some situations, much of the understorey flora of plantations comprises widespread and weedy species, including exotics (Michelsen et al, 1996). However, in other situations, plantations may contain

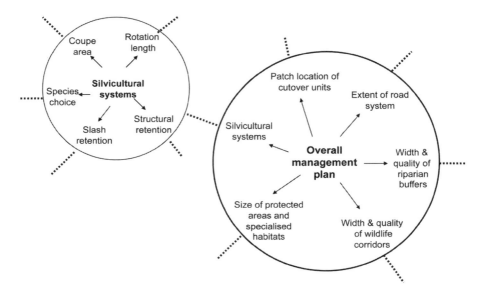

Figure 5.2 *Multi-scaled plans for forest biodiversity conservation*

Source: adapted from Lindenmayer and Franklin, 2002

a large part of the native understorey vegetation: Newmaster et al (2006) described how nearly half of the species found in comparable native forests could be found after 50 years in conifer plantations in Ontario.

However, in most cases, biodiversity in plantations cannot be restored to an 'original level' found in native forests of the same locality, regardless of the silvicultural approaches used (Bone et al, 1997; Lamb, 1998). Even where plantations are established with native species, species richness remains below that found in native forests of the same tree species, as was found for example for invertebrates in eucalypt plantations (Cunningham et al, 2005) and in teak plantations (Tangmitcharoen et al, 2006). However, silvicultural measures may enhance biodiversity, even without substantial losses in timber production rates (Hartley, 2002).

Studies dealing with silviculture and biodiversity cover a wide range of ecosystem components like different vegetation layers and taxonomic groups such as mammals, reptiles, birds and invertebrates. However, few studies have sampled a wide range of taxonomic groups at the same time and in the same place. In a large tropical landscape composed of eucalypt plantations and lowland rainforest, Barlow et al (2007) demonstrated, through intensive sampling of a range of different taxonomic groups, that these different taxonomic groups respond in very different ways to plantations and the structures within them. Similar findings were obtained from the Tumut Fragmentation Experiment (Lindenmayer and Franklin, 2002), where one group of species benefited from the mix of contrasting environments provided by exotic radiata pine plantations and remnant native eucalypt forests. A second group of species was able to persist in the landscape owing to the matrix provided by plantations, whereas the plantation matrix was inhospitable to a third group of species. In conclusion it is very difficult to interpret studies that investigated the effects of plantations on specific species or ecosystem components, especially when these studies covered only a short period of time. It is therefore not surprising that many contradictory results have been reported from such studies with limited scope (see also Bawa and Seidler, 1998).

Site preparation and residues management

The inter-rotational period is in many regards a critical time in plantation management. When most of the previous vegetation has been removed, the risk of nutrient losses is high and there is usually minimum habitat available for most species other than those dependent on disturbance. However, disturbance-dependent species (ruderal plants and highly mobile animal species adapted to open conditions) are usually not at risk in plantation landscapes. The challenge in this period is to maintain suitable conditions for those species that are adapted to later successional stages and require particular habitat structures. It is obvious that litter-inhabiting invertebrate communities and other species that depend on them such as birds or reptiles will benefit from the retention of litter and slash. At the same time, practices that conserve litter and slash will also conserve nutrients on site and are therefore beneficial to future plantation productivity (Nzila et al, 2002). However it may also increase the fire risk in plantations. The retention of coarse slash, which is more important for stand structure, will be discussed below in the context of the creation of structural complexity. The retention of coarse slash has implications for future access, and machine movement within the stand, and may therefore come at a cost incurred later in plantation rotations.

Tree species selection and mixtures

Tree species composition may also be regarded as an aspect of stand structural complexity, but because species choice is such an important management decision and because the choice of tree species in the light of vast areas of planted exotic species is so contentious, it is treated separately here. In the first part of this chapter, we have already seen that many of the beneficial effects of increased tree diversity depend on the particular combination of species with certain traits.

From a biodiversity perspective, the use of native tree species has many benefits over the use of exotics. It is generally believed that the latter harbour less diverse communities of native species, in particular invertebrates or micro-organisms (Hartley, 2002), but recent studies comparing exotic spruce plantations with native pine and oak forest in Britain have shown that this is not necessarily the case (Quine and Humphrey, 2009). Whereas the diversity and species richness of taxonomic groups such as invertebrates can be linked to the question of native vs. exotic tree species, for understorey vegetation indirect effects such as litter depth (which may restrict the germination of seeds) and microclimate (light, temperature, canopy throughfall) are more important (Parrotta, 1995; Keenan et al, 1997). Native tree species may also be important for the socio-cultural setting of the region (Lamb, 1998), particularly if they provide traditional timber or non-timber products.

There may be situations where exotic plantation species could be replaced by native species with similar properties and growth rates. The irony is that in most situations, in particular in the tropics, the knowledge base that comes with widely planted exotic trees is much better than for the native species and therefore the perceived cultivation risks are also lower for the exotics.

The question of tree species choice also relates to the choice between tree mixtures or monocultures. The merits and risks of diverse tree plantations have been covered above.

Stand density management (initial spacing and thinning)

First, we focus on the effects of stand manipulations on understorey vegetation diversity, which, in managed forests, comprises most of the floristic diversity (Halpern and Spiess, 1995). The ecosystem goods and services provided by this non-tree vegetation are manifold. They include consumptive use of plants as medicinal or edible plants or parts thereof, the use as fodder for grazing animals, as well as the ecosystem functions such as soil cover (Michelsen et al, 1996) and the sequestering and recycling of nutrients (Moore and Allen, 1999).

Any regulation of stand density has effects on canopy density, which are of variable duration depending on the stand developmental phase and the intensity of the reduction in stand density. Since plantations are mostly kept in the phase of dynamic growth associated with rapid height development, crowns can also expand rapidly and hence the canopy can potentially close within short periods following establishment or thinning.

A wide spacing of trees at the time of plantation establishment offers opportunities to support species richness. An important question is whether this effect of wider spacing lasts only until canopy closure or has longer-lasting effects on understorey species composition. In a unique study, Newmaster et al (2006) quantified species richness in conifer plantations in northern Ontario established at different initial spacing (1.8, 2.7 and 3.6m) 50–70 years after establishment. He found a clearly significant correlation between increasing spacing at planting and species richness as well as reduced woody plant abundance, higher cover of herbaceous plants and substantial increases in cryptogam cover after this time. Apparently, the duration of the early successional stage before canopy closure can influence the understorey species composition of the plantation for a long period of time. Unfortunately there are no similar studies to corroborate this observation for other forest ecosystem types.

Typically, aims of vegetation management in the establishment phase of plantations are to prevent strong competition between tree seedlings and ground vegetation, to allow rapid growth of trees, and to avoid other harmful effects of vegetation cover such as reducing herbivory damage by altering wildlife habitat. In most cases vegetation management requires some form of control of the competing vegetation, which can range from manual weeding to broadcast spraying (Evans and Turnbull, 2004). While the mid- to long-term effects to herbicide applications on vegetation may be small (Sullivan et al, 1998), any management activities that minimize the removal of existing vegetation and/or the depletion of the soil seed bank have the potential to benefit floristic diversity (Hartley, 2002). These activities include site preparation techniques that focus on the areas around planted trees, such as in spot spraying or scalping. In these instances, impacts for a given stand will decrease with increasing tree spacing (Puettmann and Berger, 2006).

In temporal sequence, the effects of thinning operations follow those of initial spacing. An often studied subject is the influence of canopy opening on the abundance and diversity of understorey vegetation. There is abundant evidence that the reduction in canopy density through thinning promotes the development of the understorey in the form of grass, herb and shrub layers (Harrington and Ewel, 1997; Bailey and Tappeiner, 1998) resulting in a positive relationship between the degree of opening and the increase in understorey biomass (Bone et al, 1997; Harrington and Edwards, 1999; Battles et al, 2001; Elliott and Knoepp, 2005). A positive relationship between vascular plant diversity and richness, and management-induced canopy openings have been found in several studies (Battles et al, 2001; Muir et al, 2002). However, there are also many studies that could not find a significant increase in floristic diversity following canopy opening in native forests (Bauhus et al, 2002; Gilliam, 2002; Elliott and Knoepp, 2005; Gondard and Romane, 2005) as well as plantations (McQueen, 1973). In addition, the effects of canopy density are not static but may change with plantation development. In particular, when plantations originate from the afforestation of former agricultural land, a phase of dense canopy may be required to reduce grasses or other dominant ruderal species to facilitate the later establishment of typical forest understorey or later successional tree species. For example, Powers et al (1997) found that establishment of woody species in the understorey was highest in *Vochysia* plantations when compared to open, abandoned fields or plantations with other dominant tree species. They attributed this facilitative effect on a period of low light conditions in the understorey, which effectively suppressed the dominant grasses and ferns, allowing other species to colonize as the plantation matured. The limitation of recruitment of typical forest understory species by grassy understoreys has also been pointed out by others (Parrotta et al, 1997). However, prolonged low light conditions beneath the canopy may limit the regeneration of all species including grasses (Powers et al, 1997). In addition, there may be situations, where the understorey is dominated by a few exotic or native species before or after thinning so that overstorey manipulations alone have little, or even negative effects on understorey species richness and diversity. This type of situation has been reported for native *Eucalyptus grandis* plantations with a dominant understorey of exotic *Lantana camara* (Cummings et al, 2007).

Different light regimes favour different understorey species but the effects of light intensity on plant species diversity may be difficult to predict. In some studies low light levels may increase species richness by suppressing dominant species, but in other cases low light may suppress the understorey altogether. If the understorey vegetation of native forests that would normally occur at the site of the plantations comprises mostly shade-tolerant species, then more intensive or frequent thinning in the plantation may even reduce the proportion of typical, native forest understorey species (Bristow et al, 2006). Therefore, there may be advantages in thinning in non-uniform patterns within stands or between stands creating a range of canopy conditions to support understorey vegetation with different light requirements.

In addition to floristic diversity, stand density management can influence the diversity of fauna in plantations. The fauna within plantations provides a range of ecosystem goods and services, which comprise the harvesting of animals for human consumption, the role of animals as antagonists for pest species, pollination and seed dispersal services, etc. (MEA, 2005b). At the same time, many species, in particular folivorous insects and mammals are considered pests and are controlled through spraying, shooting or fencing (Dolbeer et al, 1994; Speight and Wylie, 2001). The richness and diversity of animal species is related to the resources provided in plantations. The distribution of habitat resources are commonly a function of tree and understorey species density and diversity, vertical and horizontal stand structural variability, and the landscape context of plantations (Lindenmayer and Hobbs, 2004).

We are not aware of any studies that have directly investigated the effects of initial tree spacing at planting on fauna diversity in plantations. How the removal of trees through thinning affects animals in forest plantations has been studied in a few cases only and certain trends have been observed only within particular species groups. In most reported cases, insects appeared to benefit from thinning. For example, increases in ant species following thinning have been observed in the litter of *Terminalia ivornensis* plantations in Cameroon (Watt et al, 2002) and in *Pinus* plantations in Argentina (Corley et al, 2006). In open areas resulting from thinning in *Picea sitchensis* plantations in Ireland, higher spider diversity was observed when compared to dense patches (Oxbrough et al, 2006). In line-thinned Japanese cedar plantations, where between 25–35 per cent of trees were removed, the number of *Hymenoptera* families and functional groups were greater than in unthinned stands, which was related to the increased species richness and biomass of understorey plants (Maleque et al, 2007). Similar results were obtained for *Chrysomelidae* in thinned and unthinned Japanese larch plantations (Ohsawa and Nagaike, 2006), whereas another group of species, weevils, did not respond to thinning in these plantations (Ohsawa, 2005).

More individual birds and more bird species found in thinned rather than in unthinned Douglas fir stands were explained by Muir et al (2002) with the abundance of shrubs after the thinning, which caused an increase in arthropod populations (Yi and Moldenke, 2005). This trend of increasing diversity was confirmed in later studies, although here some species showed reduced population densities (Hagar et al, 2004). Similar results have been obtained by Lopez and Mori (1997) and Carey and Johnson (1995), who related the abundance and diversity of birds and small mammals, respectively, to the development of understorey and thus vertical stand structure; see also Humphrey et al (1999). However, 12–14 years after pre-commercial thinning of lodgepole pine plantations to low, medium and high residual densities, Sullivan et al (2005) could not find a significant difference in small mammal communities between these thinned stands and in comparison with unthinned and old-growth stands. For the same experiment, Ransome et al (2004) found that flying squirrel abundance was highest in dense regrowth stands, whereas there were no stand density effects on red squirrels, reiterating the point that depending on habitat requirements, species will respond differently to thinning. Also, the effect of thinning appears to take some time to manifest. Yuan et al (2005) found very little response in species richness and the abundance of small mammals and non-breeding birds to different levels of thinning in *Cryptomerica japonica* plantations.

The results above indicate that the effects of plantation thinning on fauna diversity are closely related to the creation of structural variability, which will be discussed in more depth in the next section. The effects of thinning on one trophic group (e.g. insectivorous birds) may depend primarily on the response to thinning at lower trophic levels (arthropods). To maintain or to maximize the diversity of birds and mammals at the landscape scale, a variety of thinning intensities and patterns, ranging from no thinning to widely spaced trees within and among stands may be employed (Hagar et al, 2004).

Creation of stand structural variability

Structural variability of forest stands can be regarded as a measure of the number of different structural attributes and the relative abundance of each of these attributes (McElhinny et al, 2005). As already discussed above, the focus in this section is not how to maximize structural complexity in plantations, which are commonly managed as fairly uniform stands to grow a homogeneous product. Instead, the focus is on how structural complexity may be increased without substantially compromising plantation productivity or possibly even increasing it. In almost all situations, structural complexity in plantations will be lower than in comparable native forests (Moore and Allen, 1999). The importance of structural diversity in plantations is discussed mostly with regard to fauna biodiversity since it is largely the vegetation that provides the stand structure. Therefore we will focus on the relationships between structural variability in plantations and the biodiversity of fauna and how this relationship is influenced by management.

Box 5.1 *Attributes commonly used to characterize stand structure*

Both the total or average quantity of these elements as well as their variation in space is considered important:

Stand element	Measured attribute
Foliage	foliage height diversity,[a] number of strata, foliage density
Canopy cover	canopy cover, gap size classes, gap proportion
Tree diameter	average DBH,[b] variation in DBH, diameter distribution, number of large trees
Tree height	height of overstorey, variation in tree height, height class richness
Tree spacing	different aggregation indices, number of trees per ha
Stand biomass	basal area, volume
Tree species	diversity or richness, abundance of key species
Understorey vegetation	total cover of understorey, herbaceous vegetation cover and its variation, species richness, shrub cover, shrub height, regeneration density
Dead wood	number, volume or basal area of stags, volume of coarse woody debris, log volume by decay or diameter class, variation in log density

[a] foliage height diversity is an expression of the number of vegetation strata within a vertical stand profile and the relative density of these strata (MacArthur and MacArthur, 1961).
[b] DBH = diameter at breast height

Source: after McElhinny et al, 2005

Many studies showed that plantations support biodiversity owing to the structural complexity or landscape heterogeneity (Kerr, 1999; Lanham et al, 2002; Thompson et al, 2003; Humphrey, 2005; Brown et al, 2006; Carnus et al, 2006; Lindenmayer et al, 2006 and others). Given the list of attributes in Box 5.1, it is obvious how stand structural complexity in plantations may be increased: through planting more species-diverse stands, using longer rotations, allowing understorey development, and the retention of residual trees and tree residues (see also Keeton, 2006; Bauhus et al, 2009).

The creation of more foliage layers through permitting development of the understorey provides more niches, which has already been discussed in the context of thinning responses. Two fauna groups that have been well studied with regard to vegetation structure are birds and invertebrates. Humphrey et al (1999) showed a positive correlation between vertical structure of pine and spruce stands and the diversity of syrphids (hoverflies) and carabids (ground beetles). Increasing the vertical structure of the lowest vegetation layer (0–2m) did support the diversity of carabid beetles but not of flies. Other authors who studied the diversity of insects in relation to understorey vegetation have come to similar results (Jukes et al, 2001; Ohsawa, 2005; Oxbrough et al, 2005; Ohsawa and Nagaike, 2006; Oxbrough et al, 2006).

The relationship between bird species richness and vertical stand structure has been demonstrated in the classical study by MacArthur and MacArthur (1961). In particular, increased understorey cover and species richness appear to be important to provide foraging sites. Openings in *Acacia falcataria* plantations in Borneo assisted the development of an understorey that attracted birds and increased their diversity (Mitra and Sheldon, 1993). In addition, structural variability can be created through the retention of patches of native vegetation (Hartley, 2002), sometimes called understorey islands, if they are deliberately excluded from harvesting and site preparation (Ough and Murphy, 1998). Intermittent stand-tending practices such as thinning can have strong effects on structural variability, depending on their intensity and also the technique or equipment used (e.g. cable or ground based, manual or harvester) (Figure 5.3). Loumeto and Huttel (1997) found that species richness and the proportion of forest species increased with stand age but that this development may be slowed by disturbance such as fire, herbicide treatments or other weeding practices. Likewise, thinning appeared to set back succession by increasing early successional species (McQueen, 1973).

Figure 5.3 *Disturbance of understorey vegetation through harvesting equipment in* Eucalyptus grandis *plantation, New South Wales, Australia*

Note: This disturbance could be reduced through careful planning of timber harvesting to keep the traffic of machines within stands on extraction/access tracks
Source: Photo: J. Bauhus

Cavities for nesting and dead wood as foraging sites are additional critical structural elements for many bird species and other species groups. Therefore bird diversity can be enhanced by retaining dead standing trees and also large living trees (Land et al, 1989; Kavanagh and Turner, 1994; Niemela et al, 1995). Large living trees eventually become dead wood and therefore snags, logs and other coarse woody debris in plantations. In addition, large living trees last longer than dead trees and provide cavities to those species which require such features in living trees. Dead wood on the ground, which is often pushed into windrows or piles following harvesting, is also an important structural element for birds and other fauna (Lindenmayer and Franklin, 2002). Large dead standing trees and logs are habitat for detritivores and decay organisms, provide shelter for forest-dependent vertebrates, hiding space for animals and are a long-term source of energy and nutrients. While it is generally known that retention of these structural elements (large trees, dead wood) benefits biodiversity, it is very difficult and has rarely been attempted to translate this into specific quantitative management goals for plantation forests. To advance the possible retention of structural elements in plantation forests, it is necessary to quantify the associated losses in productivity or the increase in operation costs. Bi et al (2002) and Palik et al (2003) provide examples for assessing the influence of retained mature trees on plantation productivity and seedling regeneration, respectively. The influence of these structural elements on plantation growth and operations may be reduced by not dispersing these elements over the entire area, but to keep them spatially aggregated, which is usually also easier to implement. One unresolved issue is whether aggregated retention of structural elements is as effective for habitat conservation as dispersed retention (Hartley, 2002) and what the consequences of the different patterns may be for other ecosystem structures and processes. For example, aggregated retention in Douglas fir forests caused a greater intensity of disturbance concentrated in a smaller portion of the harvested unit when compared to dispersed retention (Halpern and McKenzie, 2001).

In addition to the creation and maintenance of structural variability within plantations stands, the landscape setting and the matrix of plantation patches is also very important. At a large scale such as the landscape, spatial variability can be achieved through conserving remnant patches of native vegetation within a plantation matrix (Lamb, 1998; Lindenmayer and Franklin, 2002; Brown et al, 2006). Many studies have demonstrated the importance of patches of native vegetation for different species groups. For example, in *Pinus radiata* plantations the diversity of bird species decreased with increasing distance to native *Eucalyptus* forests and bird species richness was positively correlated with the width of natural forest strips within the plantation landscape (Tubelis et al, 2004). In an *Acacia mangium* plantation landscape on Sumatra, primates were only observed in retained patches of riparian forests that were connected to large natural forest patches in the landscape (Nasi et al, 2008). This indicates for this group of species that areas set aside from production should ideally be connected to the larger conservation areas or natural forest. The size of such patches of remnant forest within plantations will also influence the kind of species that occur. Lindenmayer and Franklin (2002) make several suggestions about how structural diversity can be maintained at the landscape level within plantation estates. This includes the maintenance of natural water bodies, riparian forests and other corridors, vegetation retention after logging through the landscape and careful planning of forest roads. It points to the importance of the basic design of plantation estates in the landscape.

Increasing rotation length

Increasing the rotation length has many different effects on the provision of habitat for biodiversity conservation. At the level of plantation stands, increasing the production cycles:

- provides more time for initial colonization and subsequent build-up of populations;
- permits the development of forest-interior microclimate;
- provides opportunities for the development of more complex stand structures; and
- reduces the frequency of harvesting-related disturbances.

Many studies have demonstrated a correlation between understorey species richness and plantation age, but this does not appear to be a linear trend. The age effect may be confounded with stand density, but this has not been demonstrated. For example, understorey species richness increased with age in pine and eucalypt plantations in South Africa (Geldenhuys, 1997). It was suggested that increased heterogeneity of light conditions as the stand ages may, at least in part, explain this relationship. In contrast, in Douglas fir plantations in Oregon, a decrease in light availability as the stand aged appeared to reduce understorey species richness (Schoonmaker and McKee, 1988). Species richness peaked at 20 years and then declined as the canopy closed over the following 10 years. A non-linear age trend of understorey diversity, following stand developmental phases, was described by Kirby (1988) for plantations on old woodland sites in Britain. Species richness of ground flora initially peaked prior to canopy closure, declined during canopy closure and the associated thicket stage and then increased after the stands thinned. Kirby (1988) suggests that this pattern is not only attributable to competition for light but also to competition for below ground resources. In other studies, late-successional species increased as plantations aged (Schoonmaker and McKee, 1988; Keenan et al, 1997; Ogden et al, 1997).

At the landscape scale, increased rotation length reduces the proportion of areas that are clear-cut or carry very young plantations. This in turn also increases the percentage of 'mature' plantation habitat and the connectivity between these patches over time. Similar results have been found for fauna as for understorey vegetation in relation to plantation age. The species richness of

Box 5.2 *Understorey plant species richness in plantations*

To assess the influence of plantation age and climatic variables on species richness of understorey vegetation, we compiled a data set from 35 published studies. Where available, the following data were entered for each study: study region, mean annual temperature, annual precipitation, plantation species, age of the plantation, shade-tolerance ranking of plantation species (tolerant, intermediate, intolerant), floristic species richness either per plot or total and the percentage of exotic understorey species. Multiple linear regression models were fitted to explain the variation in the dependent variables: total species richness, plot species richness and percentage of exotic understorey species. For this analysis it was assumed that the plot size used in each study was appropriate to measure species richness at the site in question and thus that plot species richness was comparable between studies. Annual precipitation and stand age explained 57 per cent of the variation in species richness. Similarly, annual precipitation and stand age explained 37 per cent of the variation in the natural log of plot species richness. 73 per cent of the variation in the percentage of exotic understorey species was explained by mean annual temperature, annual precipitation and stand age.

Our statistical models predicted that plot-based species richness decreased with age and total species richness would increase with age. This apparent contradiction is not surprising given that species richness does not appear to increase or decrease in a linear fashion over time but may actually peak multiple times in the life of a stand. However, if we consider plot species richness a measure of small-scale diversity and total species richness a measure of large-scale diversity, the results are not contradictory at all. It is possible that small-scale diversity decreases with time because it is more influenced by microsite heterogeneity during the stand initiation phase when competition is low. The influence of this microsite heterogeneity will decrease over time as organisms expand and outcompete their neighbours. On the other hand total species richness may be more influenced by larger-scale processes at the stand level.

Chrysomelidae (leaf beetles) was higher in old than in young Japanese larch plantations (Ohsawa and Nagaike, 2006), and Hanowski et al (1997) found that the richness of bird species communities increased with the age of poplar plantations.

Since the rotation length of plantations is usually set to optimize economic performance, which is closely linked to the culmination in mean annual increment or the culmination in value increment, an extension of rotation length will, in most cases, have negative economic consequences. Likewise the retention of other structural elements may incur costs or reduce revenue. If these measures to increase plantation biodiversity were not voluntary but were imposed on landowners, some form of compensation would be required.

Conclusions

In this chapter, we explored the interactions between biodiversity and plantations in two different ways. First, we discussed the relationships between tree species diversity of plantations and ecosystem functioning, which is the basis for the provisioning of ecosystem services. Second, we reviewed the influence of silvicultural management on the provision of the habitat function in plantations. Both parts are concerned with silvicultural choices, this includes the decision about which species to plant and whether, or how, to mix them.

The first part has shown that managing for higher tree species diversity within plantation stands has the strong potential to increase a number of ecosystem functions and hence the provision of ecosystem goods and services. However, the reasoning for this has been based on much ecological theory but on only a few concrete experiments in forest plantation systems. Here, most of the evidence comes from mixed stands of two or a few species. These show that the perceived benefits of polycultures are not being realized in every situation, and that there may also be trade-offs between the effects of mixing trees, for example achieving higher ecosystem productivity at the cost of higher water consumption. The outcomes of mixing tree species are often not diversity effects per se but depend on the particular combination of species. In addition, mixing the same combination of species may have very different outcomes depending on site and resource availability. Thus it is very difficult to predict the performance of mixed or diverse tree species plantations at a particular site without prior experimentation. This is likely to be one of the reasons for the low uptake so far of mixtures in plantation practice, despite the long and strong interest in mixed stands in the research community.

The way forward in this regard could be to follow a number of complementary approaches. On the one hand, it is necessary to develop an experimental basis to answer questions regarding tree species diversity and ecosystem functioning, as is currently under way in some parts of the world (e.g. Scherer-Lorenzen et al, 2007). On the other hand, greater research effort should be invested in identifying and quantifying the effects of functional traits of tree species on certain ecosystem functions and on their compatibility for growing in mixtures. This will reduce the risk of trying new species combinations. These points should not only be followed with regard to tree species mixtures, which were the focus here, but also for mixtures of trees with other growth forms, such as shrubs or palms that provide particular ecosystem goods (e.g. Hiremath and Ewel, 2001).

In addition, the low uptake of mixed-species plantations may also be attributable to the fact that surprisingly little of the ecological literature and body of knowledge on the relationships between biodiversity and ecosystem functioning has transgressed into the silvicultural knowledge of plantation management. This may have to do with traditions of and barriers between the disciplines (Puettmann et al, 2008). However, there are promising signs that ecologists are increasingly working on production ecosystems and also on issues related to private land (e.g. Norton, 2000) and that ecosystem functioning and related services are coming more often into the focus of silviculturists (Puettmann et al, 2008).

A further impediment to the establishment of mixed-species stands is likely to be the additional investment into the knowledge base that underpins the domestication and cultivation of tree species. The amount of basic and applied research that has been carried out to provide the knowledge base for some of the widely cultivated plantation species is enormous (e.g. Burdon, 2001). Adding new tree species to plantation systems would require a large research effort. The costs of this would have to be outweighed by the perceived benefits such as increased productivity or increased resiliency, which are of direct benefit to the plantation owner. However, additional benefits to society through improved provision of other ecosystem services should be factored into the equation. Valuing these services and more importantly the development of mechanisms to reward plantations owners for their provision might pave the way for more plantation diversity.

The second part of this chapter dealt with the habitat function of plantations and how this can be influenced by silvicultural management, mostly at the stand level. It demonstrated that the silvicultural tools and options exist to influence the provision of habitat from plantations. For the habitat function of plantation ecosystems, we know in principle how they can be manipulated, and that the methods and models exist to quantify them. The examples provided show that even intensive production systems may be valuable for the conservation of biodiversity. This has to be considered in the light of ongoing discussions about the segregation vs. the integration of ecosystem functions at the landscape level (e.g. Hartley, 2002; Sayer and Maginnis, 2005).

There are a number of ways in which plantation management can be modified to accommodate a higher level of biodiversity. However, at the operational level of plantation management, important questions still have to be addressed. First, realistic conservation goals have to be set for the particular situation. These will depend largely on the landscape context of plantations and whether species-specific conservation efforts are required. Second, for the specific local situation, it needs to be ascertained what the trade-offs are between implementing these measures and achieving the main goals of commercial plantations, or which biodiversity benefits can be strived for without impacting on the main goals of plantation management.

One very important approach for optimizing the biological value of plantation landscapes is by better landscape design and management of areas set aside for conservation. This has not been dealt with here, but has been subject of other work (e.g. Lindenmayer and Franklin, 2002). Its impact will be greatest for plantations that are to be established. Embedded in a multi-scaled approach, which considers landscape issues at a higher spatial scale, many options exist at the stand level to enhance the provision of habitat through increasing structural complexity, species mixtures, prolonged rotations and alternative site preparation. In practice, and this review has pointed to many examples, the provision of many ecosystem goods and services may be greatly enhanced at little additional cost, if an awareness of the effects of these practices has been created.

References

Alcorn, P. J., Pyttel, P., Bauhus, J., Smith, G., Thomas, D., James, R. and Nicotra, A. (2007) 'Effects of initial planting density on branch development in four-year-old plantation grown *Eucalyptus pilularis* and *E. cloeziana* trees', *Forest Ecology and Management*, vol 252, pp41–51

Anders, S., Müller, J., Augustin, S. and Rust, S. (2006) 'Die Ressource Wasser im zweischichtigen Nadel-Laub-Mischbestand', in P. Fritz (ed.) *Ökologischer Waldumbau in Deutschland: Fragen, Antworten, Perspektiven*, Oekom Verlag, pp152–183

Appasamy, P. P. (1993) 'Role of non-timber forest products in a subsistence economy: The case of a joint forestry project in India', *Economic Botany*, vol 47, pp258–267

Aretino, B., Holland, P., Peterson, D. and Schuele M. (2001) 'Creating markets for biodiversity: A case study of Earth Sanctuaries Ltd', Productivity Commission, Canberra, Australia

Assmann, E. (1970) *The Principles of Forest Yield Study: Studies in the Organic Production, Structure, Increment, and Yield of Forest Stands*, Pergamon Press, Oxford

Attiwill, P. M. (1994) 'The disturbance of forest ecosystems: The ecological basis for conservative management', *Forest Ecology and Management*, vol 63, pp247–300

Bailey, J. D. and Tappeiner, J. C. (1998) 'Effects of thinning on structural development in 40- to 100-year-old Douglas-fir stands in western Oregon', *Forest Ecology and Management*, vol 108, pp99–113

Balvanera, P., Pfisterer, A. B., Buchmann, N., He, J.-S., Nakashizuka, T., Raffaelli, D. and Schmid, B. (2006) 'Quantifying the evidence for biodiversity effects on ecosystem functioning and services', *Ecology Letters*, vol 9, pp1146-1156

Barlow, J., Gardner, T. A., Araujo, I. S., Avila-Pires, T. C., Bonaldo, A. B., Costa, J. E., Esposito, M. C., Ferreira, L. V., Hawes, J., Hernandez, M. I. M., Hoogmoed, M. S., Leite, R. N., Lo-Man-Hung, N. F., Malcolm, J. R., Martins, M. B., Mestre, L. A. M., Miranda-Santos, R., Nunes-Gutjahr, A. L., Overal,W. L., Parry, L., Peters, S. L., Ribeiro-Junior, M. A., da Silva, M. N. F., da Silva Motta, C. and Peres C. A. (2007) 'Quantifying the biodiversity value of tropical primary, secondary and plantation forests', *Proceedings of the National Academy of Science of the United States of America*, vol 104, pp18555–18560

Battles, J. J., Shlisky, A. J., Barrett, R. H., Heald, R. C. and Allen-Diaz, B. H. (2001) 'The effects of forest management on plant species diversity in a Sierran conifer forest', *Forest Ecology and Management*, vol 146, pp211–222

Bauhus, J. and Messier, C. (1999) 'Soil exploitation strategies of fine roots in different tree species of the southern boreal forest of eastern Canada', *Canadian Journal of Forest Research*, vol 29, no 2, pp260–273

Bauhus, J., Aubin, I. and Messier C. (2002) 'Composition, structure, light attenuation and nutrient content of the understorey vegetation in a *Eucalyptus sieberi* regrowth stand 6 years after thinning and fertilisation', *Forest Ecology and Management*, vol 144, pp275–286

Bauhus, J., van Winden, A. P. and Nicotra, A. B. (2004) 'Above-ground interactions and productivity in mixed-species plantations of *Acacia mearnsii* and *Eucalyptus globulus*', *Canadian Journal of Forest Research*, vol 34, pp686–694

Bauhus, J., Forrester, D. and Khanna, P. K. (2006) 'Der Einfluss von Baumartenmischungen auf Nährstoffkreisläufe und Bestandesproduktivität – Eine Fallstudie mit Mischbeständen aus *Eucalyptus globulus* and *Acacia mearnsii*', in C. Ammer, M. Schölch and R. Mosandl (eds) *Der Beitrag des Waldbaus zur Mischwaldforschung. Beiträge zur Jahrestagung der Sektion Waldbau*, Deutscher Verband Forstlicher Forschungsanstalten (DVFFA) – Freising 14–16 Sept. 2005, pp1–13

Bauhus, J., Puettmann, K. and Messier, C. (2009) 'Silviculture for old-growth attributes', *Forest Ecology and Management*, vol 258, pp525–537

Baumgärtner, S. (2007) 'The insurance value of biodiversity in the provision of ecosystem services', *Natural Resource Modeling*, vol 20, pp87–127

Bawa, K. S. and Seidler, R. (1998) 'Natural Forest management and conservation of biodiversity in tropical forests', *Conservation Biology*, vol 12, pp46–55

Berndt, L. A., Brockerhoff, E. G. and Jactel, H. (2008) 'Relevance of exotic pine plantations as a surrogate habitat for ground beetles (Carabidae) where native forest is rare', *Biodiversity and Conservation*, vol 17, pp1171–1185

Bi, H. Q., Bruskin, S. and Smith R. G. B. (2002) 'The zone of influence of paddock trees and the consequent loss in volume growth in young *Eucalyptus dunnii* plantations', *Forest Ecology and Management*, vol 165, pp305–315

Binkley, D. and Menyailo, O. (eds) (2005) *Tree Species Effects on Soils: Implications for global change*, NATO Science Series 55, Kluwer Academic Publishers, Dordrecht

Bone, R., Lawrence, M. and Mangombo, R. (1997) 'The effect of a *Eucalyptus camaldulensis* (Dehn) plantation on native woodland recovery on Ulumba Mountain, southern Malawi', *Forest Ecology and Management*, vol 99, pp83–99

Bowyer, J. L., Shmulsky, R. and Haygreen, J. C. (2003) *Forest Products and Wood Science: An Introduction*, 4th edn. Iowa State University Press, Ames, IO

Bristow, M., Nichols, J. D. and Vanclay, J. K. (2006) 'Improving productivity in mixed-species plantations – Preface', *Forest Ecology and Management*, vol 233, pp193–194

Brown, S., Palola, E. and Lorenzo, M. (2006) *The Possibility of Plantations: Integrating Ecological Forestry into Plantation Systems*, National Wildlife Federation, p46

Brundrett, M., Dell, B., Malajczuk, N. and Gong, M. Q. (1995) 'Mycorrhizas for plantation forestry in Asia', ACIAR Proceedings No. 62, Australian Centre for International Agricultural Research, Canberra

Burdon, R. D. (2001) '*Pinus radiata*', in F. T. Last (ed.) *Ecosystems of the World 19 – Tree Crop Ecosystems*, Elsevier, Amsterdam, pp99–161

Byrne, K. A. and Milne, R. (2006) 'Carbon stocks and sequestration in plantation forests in the Republic of Ireland', *Forestry*, vol 79, pp361–369

Calder, I. R. (2005) *Blue Revolution: Integrated Land and Water Resource Management*, 2nd edn, Earthscan, London

Callaway, R. M. (1995) 'Positive interactions among plants', *The Botanical Review*, vol 61, pp306–349

Carey, A. B. and Johnson, M. L. (1995) 'Small mammals in managed, naturally young, and old-growth forests', *Ecological Applications*, vol 5, pp336–352

Carnus, J. M., Parotta, J., Brockerhoff, E., Arbez, M., Jactel, H., Kremer, A., Lamb, D., O'Hara, K. and Walters, B. (2006) 'Planted forests and biodiversity', *Journal of Forestry*, vol 104, pp65–78

Caspersen, J. P. and Pacala, S. W. (2001) 'Successional diversity and forest ecosystem function', *Ecological Research*, vol 16, pp895–903

Chamberlain, J. L., Bush, R. J., Hammett, A. L. and Araman, P. A. (2002) 'Managing for non-timber products', *Journal of Forestry*, vol 100, pp8–14

Chapin III, F. S., Zavaleta, E. S., Eviner, V. T., Naylor, R. L., Vitousek, P. M., Reynolds, H. L., Hooper, D. U., Lavorel, S., Sala, O. E., Hobbie, S. E., Mack, M. C. and Díaz, S. (2000) 'Consequences of changing biodiversity', *Nature*, vol 405, pp234–242

Chen, X. (2006) 'Tree diversity, carbon storage, and soil nutrient in an old-growth forest at Changbai Mountain, Northeast China', *Communications in Soil Science and Plant Analysis*, vol 37, pp363–375

Coates, K. D. and Burton, P. J. (1997) 'A gap-based approach for development of silvicultural systems to address ecosystem management objectives', *Forest Ecology and Management*, vol 99, pp337–354.

Corley, J., Sackmann, P., Rusch, V., Bettinelli, J. and Paritsis, J. (2006) 'Effects of pine silviculture on the ant assemblages (Hymenoptera: Formicidae) of the Patagonian steppe', *Forest Ecology and Management*, vol 222, pp162–166

Cossalter, C. and Pye-Smith, C. (2003) 'Fast-wood forestry: myths and realities', Forest Perspectives, CIFOR, Jakarta

Cutter, B. E., Coggeshall, M. V., Phelps, J. E. and Stokke, D. D. (2004). 'Impacts of forest management activities on selected hardwood wood quality attributes: A review', *Wood and Fiber Science*, vol 36, pp84–97

Cummings, J., Reid, N., Davies, I. and Grant, C. (2007) 'Experimental manipulation of restoration barriers in abandoned eucalypt plantations', *Restoration Ecology*, vol 15, pp156–167

Cunningham, S. A., Floyd, R. B. and Weir, T. A. (2005) 'Do *Eucalyptus* plantations host an insect community similar to remnant *Eucalyptus* forest?' *Australian Ecology*, vol 30, no 1, pp103–117

Daily, C. D. (1997) *Nature's Services – Societal Dependence on Natural Ecosystems*, Island Press, Washington

de Groot, R. S. (1992) *Functions of Nature, Evaluation of Nature in Environmental Planning, Management and Decision Making*, Wolters-Noordhoff, Groningen

de Groot, R. S., Wilson, M. and Boumans, R. (2002) 'A typology for the description, classification and valuation of ecosystem functions, goods and services' (pp393–408). in 'The dynamics and value of ecosystem services: Integrating economic and ecological perspectives', *Ecological Economics*, vol 41, pp367–567

Dhôte, J.-F. (2005) 'Implications of forest diversity in resistance to strong winds', in M. Scherer-Lorenzen, C. Körner and E. D. Schulze (eds) *Forest Diversity and Function: Temperate and Boreal Systems*, Ecological Studies 176, Springer, Berlin, pp291–307

Dolbeer, R. A., Holler, N. R. and Hawthorn, D. W. (1994) 'Identification and control of wildlife damage', in T. A. Bookhout (ed.) *Research and Management Techniques for Wildlife and Habitats*, The Wildlife Society, Bethesda, MD, pp474–506

Elliott, K. J. and Knoepp, J. D. (2005) 'The effects of three regeneration harvest methods on plant diversity and soil characteristics in the Southern Appalachians', *Forest Ecology and Management*, vol 211, pp296–317

Emery, M., Martin, S. and Dyke, A. (2006). *Wild Harvests from Scottish Woodlands: Social, Cultural and Economic Values of Contemporary Non-timber Forest Products*, Forestry Commission, Edinburgh

Evans, J. and Turnbull, J. W. (2004) *Plantation Forestry in the Tropics*, 3rd edn, Oxford University Press, Oxford

Ewel, J. J. (1986) 'Designing agroecosystems for the humid tropics', *Annual Review of Ecology and Systematics*, vol 17, pp245–271

Facelli, J. M. and Pickett, S. T. A. (1991) 'Plant litter: Its dynamics and effects on plant community structure', *The Botanical Review*, vol 57, pp1–32

FAO (2005) '*Global Forest Resources Assessment 2005*, FAO Forestry Paper 147, FAO, Rome

Fearnside, P. M. (1999) 'Biodiversity as an environmental service in Brazil's Amazonian forests: Risks, value and conservation', *Environmental Conservation*, vol 26, pp305–321

Fernandes, P. M., Vega, J. A., Jiménez, E. and Rigolot, E. (2008) 'Fire resistance of European pines', *Forest Ecology and Management*, vol 256, pp246–255

Forrester, D., Bauhus, J. and Khanna, P. K. (2004) 'Growth dynamics in a mixed species plantation of *Eucalyptus globulus* and *Acacia mearnsii*' *Forest Ecology and Management*, vol 193, pp81–95

Forrester, D. I., Bauhus, J. and Cowie, A. L. (2005). 'Nutrient cycling in a mixed-species plantation of *Eucalyptus globulus* and *Acacia mearnsii*', *Canadian Journal of Forest Research*, vol 35, pp2942–2950

Forrester, D. I., Bauhus, J., Cowie, A. L. and Vanclay, J. K. (2006) 'Mixed-species plantations of *Eucalyptus* with nitrogen-fixing trees: A review', *Forest Ecology and Management*, vol 233, pp211–230

Forrester, D. I., Theiveyanathan, S., Collopy, J. J. and Marcar, N. E. (2010) 'Enhanced water use efficiency in a mixed *Eucalyptus globulus* and *Acacia mearnsii* plantation', *Forest Ecology and Management*, vol, 259, pp1761–1770

Foster, D. R. (1988) 'Species and stand response to catastrophic wind in Central New England, USA', *Journal of Ecology*, vol 76, pp135–151

Gartner, T. B. and Cardon, Z. G. (2004) 'Decomposition dynamics in mixed-species leaf litter', *Oikos*, vol 104, pp230–246

Gautam, K. H. (2001) 'Lopping regimes in community-managed sal (*Shorea robusta*) forests of Nepal: Prospects for multiple-product silviculture for community foresty', School of Forestry, University of Canterbury Christchurch, Canterbury

Gautam, K. H. and Devoe, N. N. (2006) 'Ecological and anthropogenic niches of sal (*Shorea robusta* Gaertn. f.) forest and prospects for multiple-product forest management – a review', *Forestry*, vol 79, pp81–101

Geldenhuys, C. J. (1997) 'Native forest regeneration in pine and eucalypt plantations in northern province, South Africa', *Forest Ecology and Management*, vol 99, pp101–115

Giardina, C. P., Ryan, M. G., Hubbard, R. M. and Binkley, D. (2001) 'Tree species and soil textural controls on carbon and nitrogen mineralization rates', *Soil Science Society of America Journal*, vol 65, pp1272–1279

Gilliam, F. S. (2002) 'Effects of harvesting on herbaceous layer diversity of a central Appalachian hardwood forest in West Virginia, USA', *Forest Ecology and Management*, vol 155, pp33–43

Gleixner, G., Kramer, C., Hahn, V. and Sachse, D. (2005) 'The effect of biodiversity on carbon storage in soils', in M. Scherer-Lorenzen, C. Körner and E. D. Schulze (eds) *Forest Diversity and Function: Temperate and Boreal Systems*, Ecological Studies 176, Springer, Berlin, pp165–183

Gonçalves, J. L. M., Barros, N. F., Nambiar, E. K. S. and Novais, R. F. (1997) 'Soil and stand management for short-rotation plantations', in E. K. S. Nambiar and A. G. Brown (eds) *Management of Soil, Nutrients and Water in Tropical Plantation Forests*, ACIAR, Canberra, Monograph No. 43, pp379–418

Gondard, H. and Romane, F. (2005) 'Long-term evolution of understorey plant species composition after logging in chestnut coppice stands (Cevennes Mountains, southern France)', *Annals of Forest Science*, vol 62, pp333–342

Hagar, J., Howlin, S. and Ganio, L. (2004) 'Short-term response of songbirds to experimental thinning of young Douglas-fir forests in the Oregon Cascades', *Forest Ecology and Management*, vol 199, pp333–347

Halpern, C. B. and Spiess, T. A. (1995) 'Plant species diversity in natural and managed forests of the Pacific Northwest', *Ecological Applications*, vol 5, pp913–934

Halpern, C. B. and McKenzie, D. (2001) 'Disturbance and post-harvest ground conditions in a structural retention experiment', *Forest Ecology and Management*, vol 154, pp215–225

Hanowski, J. M., Niemi, G. J. and Christian, D. C. (1997) 'Influence of within-plantation heterogeneity and surrounding landscape composition on avian communities in hybrid poplar plantations', *Conservation Biology*, vol 11, pp936–944

Harrington, R. A. and Ewel, J. J. (1997) 'Invasibility of tree plantations by native and non-indigenous plant species in Hawaii', *Forest Ecology and Management*, vol 99, pp153–162

Harrington, T. B. and Edwards, M. B. (1999) 'Understory vegetation, resource availability, and litterfall responses to pine thinning and woody vegetation control in longleaf pine plantations', *Canadian Journal of Forest Research*, vol 29, pp1055–1064

Hartley, M. J. (2002) 'Rationale and methods for conserving biodiversity in plantation forests', *Forest Ecology and Management*, vol 155, pp81–95

Hättenschwiler, S. (2005) 'Effects of tree species diversity on litter quality and decomposition', in M. Scherer-Lorenzen, C. Körner and E. D. Schulze (eds) *Forest Diversity and Function: Temperate and Boreal Systems*, Ecological Studies 176, Springer, Berlin, pp149–164

Hein, S., Weiskittel, A. R. and Kohnle, U. (2008) 'Effect of wide spacing on tree growth, branch and sapwood properties of young Douglas-fir in south-western Germany', *European Journal of Forest Research*, vol 127, pp481–493

Hillebrand, H. and Matthiessen, B. (2009) 'Biodiversity in a complex world: Consolidation and progress in functional biodiversity research', *Ecology Letters*, vol 12, pp1405–1419

Hiremath, A. J. and Ewel, J. J. (2001) 'Ecosystem nutrient use efficiency, productivity, and nutrient accrual in model tropical communities', *Ecosystems*, vol 4, pp669–682

Hooper, D. U., Chapin, F. S. I., Ewel, J. J., Hector, A., Inchausti, P., Lavorel, S., Lawton, J. H., Lodge, D., Loreau, M., Naeem, S., Schmid, B., Setälä, H., Symstad, A. J., Vandermeer, J. and Wardle, D. A. (2005) 'Effects of biodiversity on ecosystem functioning: A consensus of current knowledge and needs for future research', *Ecological Monographs*, vol 75, pp3–36.

Humphrey, J. W. (2005) 'Benefits to biodiversity from developing old-growth conditions in British upland spruce plantations: A review and recommendations', *Forestry*, vol 78, pp33–53

Humphrey, J. W., Hawes, C., Peace, A. J., Ferris-Kaan, R. and Jukes, M. R. (1999) 'Relationships between insect diversity and habitat characteristics in plantation forests', *Forest Ecology and Management*, vol 113, pp11–21

Jactel, H. and Brockerhoff, E. G. (2007) 'Tree diversity reduces herbivory by forest insects', *Ecology Letters*, vol 10, pp835–848

Jactel, H., Brockerhoff, E. and Duelli, P. (2005) 'A test of the biodiversity–stability theory: Meta-analysis of tree species diversity effects on insect pest infestations, and re-examination of responsible factors', in M. Scherer-Lorenzen, C. Körner and E. D. Schulze (eds) *Forest Diversity and Function: Temperate and Boreal Systems*, Ecological Studies 176, Springer, Berlin, pp235–262

Johnson, D. W. (1992). 'Effects of forest management on soil carbon storage', *Water, Air, and Soil Pollution*, vol 64, pp83–120

Johnson, V. J. (1975) 'Hardwood fuel-breaks for north eastern United States', *Journal of Forestry*, vol 73, pp588–589.

Jones, M. and Harper, J. L. (1987) 'The influence of neighbours on the growth of trees: II. The fate of buds on long and short shoots in *Betula pendula*', *Proceedings of the Royal Society of London Series B – Biological Sciences*, vol 232, pp19–33

Jukes, M. R., Peace, A. J. and Ferris, R. (2001) 'Carabid beetle communities associated with coniferous plantations in Britain: The influence of site, ground vegetation and stand structure', *Forest Ecology and Management*, vol 148, pp271–286

Kaushal, K. K. and Melkani, V. K. (2005) 'India: Achieving the Millennium Development Goals through non-timber forest products', *International Forestry Review*, vol 7, pp128–134

Kavanagh, R. P. and Turner, R. J. (1994) 'Birds in eucalypt plantations: The likely role of retained habitat trees', *Australian Birds*, vol 28, pp32–41

Keenan, R., Lamb, D., Woldring, O., Irvine, T. and Jensen, R. (1997) 'Restoration of plant biodiversity beneath tropical tree plantations in Northern Australia', *Forest Ecology and Management*, vol 99, pp117–131

Keeton, W. S. (2006) 'Managing for late-successional/old-growth forest characteristics in northern hardwood-conifer forests', *Forest Ecology and Management*, vol 235, pp129–142

Kelty, M. J. (1992) 'Comparative productivity of monocultures and mixed-species stands', in M. J. Kelty, B. C. Larson and C. D. Oliver (eds) *The Ecology and Silviculture of Mixed-Species Forests*, Kluwer Academic Publishers, Dordrecht, Netherlands, pp125–141

Kelty, M. J. (2006). 'The role of species mixtures in plantation forestry', *Forest Ecology and Management*, vol 233, pp195–204

Kelty, M. J. and Cameron, I. R. (1995) 'Plot designs for the analysis of species interactions in mixed stands', *Commonwealth Forestry Review*, vol 74, pp322–332

Kerr, G. (1999) 'The use of silvicultural systems to enhance the biological diversity of plantation forests in Britain', *Forestry*, vol 72, pp191–205

Kirby, K. J. (1988) 'Changes in the ground flora under plantations on ancient woodland sites', *Forestry*, vol 61, pp317–338

Kirschenmann, F. L. (2007) 'Potential for a new generation of biodiversity in agroecosystems of the future', *Agronomy Journal*, vol 99, pp373–376

Klooster, D. and Masera, O. (2000) 'Community forest management in Mexico: Carbon mitigation and biodiversity conservation through rural development', *Global Environmental Change*, vol 10, pp259–272

Koricheva, J., Vehviläinen, H., Riihimäki, J., Ruohomäki, K., Kaitaniemi, P. and Ranta, H. (2006) 'Diversification of tree stands as a means to manage pests and diseases in boreal forests: Myth or reality?' *Canadian Journal of Forest Research*, vol 36, pp324–336

Kumar, B. M. and Nair, P. K. R. (eds) (2006) *Tropical Homegardens: A Time-tested Example of Sustainable Agroforestry*, Springer, Dordrecht

Laclau, J. P., Bouillet, J. P., Gonçalves, J. L. M., Silva, E. V., Jourdan, C., Cunha, M. C. S., Moreira, M. R., Saint-André, L., Gonçalves, M. R., Maquère, V., Nouvellon, Y. and Ranger, J. (2008) 'Mixed-species plantations of *Acacia mangium* and *Eucalyptus grandis* in Brazil: Growth dynamics and net primary production', *Forest Ecology and Management*, vol 255, pp3905–3917

Lamb, D. (1998) 'Large-scale ecological restoration of degraded tropical forest lands: The potential role of timber plantations', *Restoration Ecology*, vol 6, pp271–279

Land, D., Marion, W. R. and Omeara, T. E. (1989) 'Snag availability and cavity nesting birds in slash pine plantations', *Journal of Wildlife Management*, vol 53, pp1165–1171

Lanham, J. D., Keyser, P. D., Brose, P. H. and van Lear, D. H. (2002) 'Oak regeneration using the shelterwood-burn technique: Management options and implications for songbird conservation in the southeastern United States', *Forest Ecology and Management*, vol 155, pp143–152

Last, F. T. (ed.) (2001) *Tree Crop Ecosystems*, Ecosystems of the World 19, Elsevier, Amsterdam

Law, B. E., Falge, E., Guc, L., Baldocchi, D. D., Bakwind, P., Berbigier, P., Davis, K., Dolmang, A. J., Falk, M., Fuentes, J. D., Goldstein, A., Granier, A., Grelle, A., Hollinger, D., Janssens, I. A., Jarvis, P., Jensen, N. O., Katul, G., Mahli, Y., Matteucci, G., Meyers, T., Monsont, R., Munger, W., Oechel, W., Olson, R., Pilegaard, K., Paw, K. T., Thorgeirsson, H., Valentini, R., Verma, S., Vesala, T., Wilson, K. and Wofsy, S. (2002) 'Environmental controls over carbon dioxide and water vapor exchange of terrestrial vegetation', *Agricultural and Forest Meteorology*, vol 113, pp97–120

Lindenmayer, B. and Franklin, J. F. (2002) *Conserving Forest Biodiversity: A Comprehensive Multiscaled Approach*, Island Press, Washington, Covelo, London

Lindenmayer, D. B. and Hobbs, R. J. (2004) 'Fauna conservation in Australian plantation forests – a review', *Biological Conservation*, vol 119, pp151–168

Lindenmayer, D. B., Franklin, J. F. and Fischer, J. (2006) 'General management principles and a checklist of strategies to guide forest biodiversity conservation', *Biological Conservation*, vol 131, pp433–445

Lopez, G. and Mori, M. J. (1997) 'Birds of Aleppo pine plantations in South-East Spain in relation to vegetation composition and structure', *Journal of Applied Ecology*, vol 34, pp1257–1272

Loreau, M. and Hector, A. (2001) 'Partitioning selection and complementarity in biodiversity experiments', *Nature*, vol 412, pp72–76

Loreau, M., Naeem, S. and Inchausti, P. (eds) (2002) *Biodiversity and Ecosystem Functioning*, Oxford University Press, New York

Loumeto, J. J. and Huttel, C. (1997) 'Understory vegetation in fast-growing tree plantations on savanna soils in Congo', *Forest Ecology and Management*, vol 99, pp65–81

Lüpke von, B. and Spellmann, H. (1997) 'Aspekte der Stabilität und des Wachstums von Mischbeständen aus Fichte und Buche als Grundlage für waldbauliche Entscheidungen', *Forstarchiv*, vol 68, pp167–179

Lugo, A. E. (1997) 'The apparent paradox of reestablishing species richness on degraded land with monocultures', *Forest Ecology and Management*, vol 99, pp9–19

Lynch, K. A. and McLain, J. (2004) 'Non-timber forest product inventorying and monitoring in the United States: Rationale and recommendations for a participatory approach', Institute for Culture and Ecology, www.ifcae.org

MacArthur, R. H. and MacArthur, J. W. (1961) 'On bird species diversity', *Ecology*, vol 42, pp594–598

McElhinny, C., Gibbons, P., Brack, C. and Bauhus, J. (2005) 'Forest and woodland stand structural complexity: Its definition and measurement', *Forest Ecology and*

Management, vol 218, pp1–24

McQueen, D. R. (1973) 'Changes in understorey vegetation and fine root quantity following thinning of 30-year *Pinus radiata* in central North Island, New Zealand', *Journal of Applied Ecology*, vol 10, pp13–21

Mahapatra, A. and Mitchell, C. P. (1997) 'Sustainable development of non-timber forest products: Implication for forest management in India', *Forest Ecology and Management*, vol 94, pp15–29

Maleque, M. A., Ishii, H. T., Maeto, K. and Taniguchi, S. (2007) 'Line thinning fosters the abundance and diversity of understory Hymenoptera (Insecta) in Japanese cedar (*Cryptomeria japonica* D. Don) plantations', *Journal of Forest Research*, vol 12, pp14–23

Maraseni, T. N. (2008) 'Selection of non-timber forest species for community and private plantations in the high and low altitude areas of Makawanpur District, Nepal', *Small-scale Forestry*, vol 7, pp151–161

MEA (Millennium Ecosystem Assessment) (2005a) *Living Beyond Our Means – Natural Assets and Human Well-being*, statement from the board, Washington, DC

MEA (2005b) *Ecosystems and Human Well-being, Synthesis*, Island Press, Washington, DC

Michelsen, A., Lisanework, N., Friis, I. and Holst N. (1996) 'Comparisons of understorey vegetation and soil fertility in plantations and adjacent natural forests in the Ethiopian highlands', *Journal of Applied Ecology*, vol 33, pp627–642

Mitra, S. and Sheldon, F. H. (1993) 'Use of an exotic tree plantation by Bornean lowland forest birds', *Auk*, vol 110, pp529–540

Montagu, K. D., Kearney, D. E. and Smith, R. G. B. (2003) 'The biology and silviculture of pruning planted eucalypts for clear wood production – a review', *Forest Ecology and Management*, vol 179, pp1–13

Montes, F., Sanchez, M., del Rio, M. and Canellas I. (2005) 'Using historic management records to characterize the effects of management on the structural diversity of forests', *Forest Ecology and Management*, vol 207, pp279–293

Moore, S. E. and Allen, H. L. (1999) 'Plantation forestry', in M. L. J. Hunter (ed.) *Maintaining Biodiversity in Forest Ecosystems*, Cambridge University Press, Cambridge, pp400–433

Muir, P. S., Mattingly, R. L., Tappeiner, II, J. C., Bailey, J. D., Elliott, W. E., Hagar, J. C., Miller, J. C., Peterson, E. B. and Starkey, E. E. (2002) *Managing for Biodiversity in Young Douglas-Fir Forests of Western Oregon*, National Technical Information Service, Springfield, Virginia

Naeem, S. (2002) 'Ecosystem consequences of biodiversity loss: The evolution of a paradigm', *Ecology*, vol 83, pp1537–1552

Naeem, S., Loreau, M. and Inchausti, P. (2002) 'Biodiversity and ecosystem functioning: The emergence of a synthetic ecological framework' in M. Loreau, S. Naeem and P. Inchausti (eds) *Biodiversity and Ecosystem Functioning: Synthesis and Perspectives* Oxford University Press, Oxford and New York, pp3–11

Nair, K. S. S. (2001) *Pest Outbreaks in Tropical Forest Plantations – Is There a Greater Risk for Exotic Trees Species?* CIFOR, Indonesia

Nasi, R., Koponen, P., Poulsen, J. G., Buitenzorgy, M. and Rusmantoro, W. (2008) 'Impact of landscape and corridor design on primates in a large-scale industrial tropical plantation landscape', *Biodiversity Conservation*, vol 17, pp1105–1126

Neumann, R. P. and Hirsch, E. (2000) *Commercialisation of Non-Timber Forest PRODUCTS: Review and Analysis of Research*, CIFOR, Bogor, Indonesia

Newmaster, S. G., Bell, F. W., Roosenboom, C. R., Cole, H. A. and Towill, W. D.

(2006) 'Restoration of floral diversity through plantations on abandoned agricultural land', *Canadian Journal of Forest Research*, vol 36, pp1218–1235

Niemela, T., Renvall, P. and Penttila, R. (1995) 'Interactions of fungi at late stages of wood decomposition', *Annales Botanici Fennici*, vol 32, pp141–152

Norton, D. A. (1998) 'Indigenous biodiverstiy conservation and plantation forestry: Options for the future', *New Zealand Forestry*, vol 43, pp34–39

Norton, D. A. (2000) 'Editorial: Conservation biology and private land: Shifting the focus', *Conservation Biology*, vol 14, pp1221–1223

Noss, R. F. (1990) 'Indicators for monitoring biodiversity: A hierarchical approach', *Conservation Biology*, vol 4, pp355–364

Nyland, R. D. (2002) *Silviculture: Concepts and Applications*, 2nd edn, The McGraw-Hill Companies, Inc., New York

Nzila, J. D., Bouillet, J. P., Laclau, J. P. and Ranger, J. (2002) 'The effects of slash management on nutrient cycling and tree growth in *Eucalyptus* plantations in the Congo', *Forest Ecology and Management*, vol 171, pp209–221

Ogden, J., Braggins, J., Stretton, K. and Anderson, S. (1997) 'Plant species richness under *Pinus radiata* stands on the central North Island Volcanic Plateau, New Zealand', *New Zealand Journal of Ecology*, vol 21, pp17–29

Ohsawa, M. (2005) 'Species richness and composition of Curculionidae (Coleoptera) in a conifer plantation, secondary forest, and old-growth forest in the central mountainous region of Japan', *Ecological Research*, vol 20, pp632–645

Ohsawa, M. and Nagaike, T. (2006) 'Influence of forest types and effects of forestry activities on species richness and composition of Chrysomelidae in the central mountainous region of Japan', *Biodiversity and Conservation*, vol 15, pp1179–1191

Ormeno, E., Cespedes, B., Sanchez, I. A., Velasco-Garcia, A., Moreno, J. M., Fernandez, C. and Baldy, V. (2009) 'The relationship between terpenes and flammability of leaf litter', *Forest Ecology and Management*, vol 257, pp471–482

Ough, K. and Murphy, A. (1998) 'Understorey islands: A method of protecting understorey flora during clearfelling operations', VSP Internal Report 29, Department of Natural Resources and Environment, Victoria

Oxbrough, A. G., Gittings, T., O'Halloran, J., Giller, P. S. and Smith, G. F. (2005) 'Structural indicators of spider communities across the forest plantation cycle', *Forest Ecology and Management*, vol 212, pp171–183

Oxbrough, A. G., Gittings, T., O'Halloran, J., Giller, P. S. and Kelly, T. C. (2006) 'The influence of open space on ground-dwelling spider assemblages within plantation forests', *Forest Ecology and Management*, vol 237, pp404–417

Palik, B., Mitchell, R. J., Pecot, S., Battaglia, M. and Pu, M. (2003) 'Spatial distribution of overstory retention influences resources and growth of longleaf pine seedlings', *Ecological Applications*, vol 13, pp674–686

Panayotou, T. and Ashton, P. S. (1992) *Not by Timber Alone: Economics and Ecology for Sustaining Tropical Forests*, Island Press, Washington, DC and Covelo, California

Parrotta, J. A. (1995) 'Influence of overstory composition on understory colonization by native species in plantations on a degraded tropical site', *Journal of Vegetation Science*, vol 6, pp627–636

Parrotta, J. A. (1999) 'Productivity, nutrient cycling, and succession in single- and mixed-species plantations of *Casuarina equisetifolia*, *Eucalyptus robusta*, and *Leucaena leucocephala* in Puerto Rico', *Forest Ecology and Management*, vol 124, pp45–77

Parrotta, J. A., Knowles, O. H. and Wunderle, J. M. (1997) 'Development of floristic diversity in 10-year-old restoration forests on a bauxite mined site in Amazonia', *Forest Ecology and Management*, vol 99, pp21–42

Parry, L., Barlow, J. and Peres, C. A. (2009) 'Allocation of hunting effort by Amazonian smallholders: Implications for conserving wildlife in mixed-use landscapes', *Biological Conservation*, vol 142, pp1777–1786

Piotto, D. (2008) 'A meta-analysis comparing tree growth in monocultures and mixed plantations', *Forest Ecology and Management*, vol 255, pp781–786

Powers, J. S., Haggar, J. P. and Fisher, R. F. (1997) 'The effect of overstory composition on understory wood regeneration and species richness in 7-year-old plantations in Costa Rica', *Forest Ecology and Management*, vol 99, pp43–54

Pretzsch, H. (2005) 'Diversity and productivity in forests: Evidence from long-term experimental plots', in M. Scherer-Lorenzen, C. Körner and E. D. Schulze (eds) *Forest Diversity and Function: Temperate and Boreal Systems*, Ecological Studies 176, Springer, Berlin, Heidelberg, New York, pp41–64

Puettmann, K. J. and Berger, C. (2006) 'Development of tree and understory vegetation in young Douglas-fir plantations in western Oregon', *Western Journal of Applied Forestry*, vol 21, pp94–101

Puettmann, K. J., Coates, K. D. and Messier, C. (2008) *Silviculture: Managing Complexity*, Island Press, Washington, DC

Quine, C. P. and Humphrey, J. W. (2009) 'Plantations of exotic tree species in Britain: Irrelevant for biodiversity or novel habitat for native species?' *Biodiversity and Conservation* online, doi 10.1007/s10531-009-9771-7

Ransome, D. B., Lindgren, P. M. F., Sullivan, D. S. and Sullivan, T. P. (2004) 'Long-term responses of ecosystem components to stand thinning in young lodgepole pine forest: 1. Population dynamics of northern flying squirrels and red squirrels', *Forest Ecology and Management*, vol 202, pp355–367

Redondo-Brenes, A. and Montagnini, F. (2006) 'Growth, productivity, aboveground biomass, and carbon sequestration of pure and mixed tree plantations in the Caribbean lowlands of Costa Rica', *Forest Ecology and Management*, vol 232, pp168–178

Resh, S. C., Binkley, D. and Parrotta, J. A. (2002) 'Greater soil carbon sequestration under nitrogen-fixing trees compared with *Eucalyptus* species', *Ecosystems*, vol 5, pp217–231

Richards, A. E., Forrester, D. I., Bauhus, J. and Scherer-Lorenzen, M. (2010) 'Physiological changes within species influence the nutrition and productivity of mixed species tree plantations: A review', *Tree Physiology* (in review)

Robinson, J. G. and Bennett, E. L. (2004) 'Having your wildlife and eating it too: An analysis of hunting sustainability across tropical ecosystems', *Animal Conservation*, vol 7, pp397–408

Rothe, A. and Binkley, D. (2001) 'Nutritional interactions in mixed species forests: A synthesis', *Canadian Journal of Forest Research*, vol 31, pp1855–1870

Saha, K. S., Nair, P. K. R., Nair, V. D. and Kumar, B. M. (2009) 'Soil carbon stock in relation to plant diversity of homegardens in Kerala, India', *Agroforestry Systems*, vol 76, pp53–65

Sayer, J. A. and Maginnis, S. (eds) (2005) *Forests in Landscapes: Ecosystem Approaches to Sustainability*, Earthscan, London

Scarff, F. R. and Westoby, M. (2006) 'Leaf litter flammability in some semi-arid Australian woodlands', *Functional Ecology*, vol 20, pp745–752

SCBD (Secretariat of the Convention on Biological Diversity) (2001) 'Sustainable management of non-timber forest resources', in SCBD (ed.) CBD Technical Series No. 6, Montreal

Scherer-Lorenzen, M., Körner, C. and Schulze, E. D. (eds) (2005) *Forest Diversity and Function: Temperate and Boreal Systems*, Ecological Studies 176, Springer, Berlin

Scherer-Lorenzen, M., Schulze, E. D., Don, A., Schumacher, J. and Weller, E. (2007) 'Exploring the functional significance of forest diversity: A new long-term experiment with temperate tree species (BIOTREE)', *Perspectives in Plant Ecology, Evolution and Systematics*, vol 9, pp53–70

Scherr, S., White, A. and Khare, A. (eds) (2004) *For Services Rendered – The Current Status and Future Potential of Markets for the Ecosystem Services Provided by Tropical Forests*, ITTO Technical Series, 21, International Tropical Timber Organization, Yokohama, Japan

Schmid, I. and Kazda, M. (2002) 'Root distribution of Norway spruce in monospecific and mixed stands on different soils', *Forest Ecology and Management*, vol 159, pp37–47

Schoonmaker, P. and McKee, A. (1988) 'Species composition and diversity during secondary succession of coniferous forests in the western cascade mountains of Oregon', *Forest Science*, vol 34, pp960–979

Schroth, G. and da Mota, M. S. S. (2004) 'The role of agroforestry in biodiversity conservation in the tropics: A synthesis of evidence', Paper presented at the International Ecoagriculture Conference and Practitioners' Fair, held in Nairobi, Kenya, The World Agroforestry Centre (ICRAF), Nairobi, September–October

Schulze, E. D., Chapin, I. F. S. and Gebauer, G. (1994) 'Nitrogen nutrition and isotope differences among life forms at the northern tree line of Alaska', *Oecologia*, vol 100, pp406–41

Schume, H., Jost, G. and Hager, H. (2004) 'Soil water depletion and recharge patterns in mixed and pure forest stands of European beech and Norway spruce', *Journal of Hydrology*, vol 289, pp258–274.

Schütz, J. P., Gotz, M., Schmid, W. and Mandallaz, D. (2006). 'Vulnerability of spruce (*Picea abies*) and beech (*Fagus sylvatica*) forest stands to storms and consequences for silviculture', *European Journal of Forest Research*, vol 125, pp291–302

Scott, D. F., Bruijnzeel, L. A. and Mackensen, J. (2005) 'The hydrological and soil impacts of forestation in the tropics', in M. Bonnell and L. A. Bruijnzeel (eds) *Forests, Water and People in the Humid Tropics*, Cambridge University Press, Cambridge, pp622–651

Sedjo, R. A. (1999) 'The potential of high-yield plantation forestry for meeting timber needs', *New Forests*, vol 17, pp339–360.

Seeland, K. (1999) 'Recent developments in social and community forestry in India, Nepal and Bhutan', in *Proceedings of Community Forestry, A Change for the Better*, London, pp32–42

Spehn, E. M., Hector, A., Joshi, J., Scherer-Lorenzen, M., Schmid, B., Bazeley-White, E. Beierkuhnlein, C., Caldeira, M. C., Diemer, M., Dimitrakopoulos, P. G., Finn, J. A., Freitas, H., Giller, P. S., Good, J., Harris, R., Högberg, P., Huss-Danell, K., Jumpponen, A., Koricheva, J., Leadley, P. W., Loreau, M., Minns, A., Mulder, C. P. H., O'Donovan, G., Otway, S. J., Palmborg, C., Pereira, J. S., Pfisterer, A. B., Prinz, A., Read, D. J., Schulze, E.-D., Siamantziouras, A.-S. D., Terry, A. C., Troumbis, A. Y., Woodward, F. I., Yachi, S. and Lawton, J. H. (2005) 'Ecosystem effects of biodiversity manipulations in European grasslands', *Ecological Monographs*, vol 75, pp37–63

Speight, M. R. and Wylie, F. R. (2001) *Insect Pests in Tropical Forestry*, CABI Publishing, Wallingford

Spiecker, H., Hansen, J., Klimo, E., Skovsgaard, J. P., Sterba, H. and Teuffel von, K. (2004) *Norway Spruce Conversion – Options and Consequences*, EFI Research Report 18, Leiden, Boston

Sullivan, T. P., Wagner, R. G., Pitt, D. G., Lautenschlager, R. A. and Chen, D. G. (1998) 'Changes in diversity of plant and small mammal communities after herbicide application in sub-boreal spruce forest', *Canadian Journal of Forest Research*, vol 28, pp168–177

Sullivan, T. P., Sullivan, D. S., Lindgren, P. M. F and Ransome, D. B. (2005) 'Long-term responses of ecosystem components to stand thinning in young lodgepole pine forest: II. Diversity and population dynamics of forest floor small mammals', *Forest Ecology and Management*, vol 205, pp1–14

Sumida, A., Terazawa, I., Togashi, A. and Komiyama, A. (2002) 'Spatial arrangement of branches in relation to slope and neighbourhood competition', *Annals of Botany*, vol 89, pp301–310

Talkner, U., Jansen, M. and Beese, F. O. (2009) 'Soil phosphorus status and turnover in central-European beech forest ecosystems with differing tree species diversity', *European Journal of Soil Science*, vol 60, pp338–346

Tangmitcharoen, S., Takaso, T., Siripatanadilox, S., Tasen, W. and Owens, J. N. (2006) 'Insect biodiversity in flowering teak (*Tectona grandis* L.f.) canopies: Comparison of wild and plantation stands', *Forest Ecology and Management*, vol 222, pp99–107

Thompson, I. D., Baker, J. A. and Ter-Mikaelian, M. (2003) 'A review of the long-term effects of post-harvest silviculture on vertebrate wildlife, and predictive models, with an emphasis on boreal forests in Ontario, Canada', *Forest Ecology and Management*, vol 177, pp441–469

Ticktin, T. (2004) 'The ecological implications of harvesting non-timber forest products', *Journal of Applied Ecology*, vol 41, pp11–21

Ticktin, T., Nantel, P., Ramirez, F. and Johns, T. (2002) 'Effects of variation on harvest limits for nontimber forest species in Mexico', *Conservation Biology*, vol 16, pp691–705

Tilman, D., Knops, J., Wedin, D., Reich, P., Ritchie, M. and Siemann, E. (1997) 'The influence of functional diversity and composition on ecosystem processes', *Science*, vol 277, pp1300–1302

Trauernicht, C. and Ticktin, T. (2005) 'The effects of non-timber forest product cultivation on the plant community structure and composition of a humid tropical forest in southern Mexico', *Forest Ecology and Management*, vol 219, pp269–278

Tubelis, D. P., Lindenmayer, D. B. and Cowling A. (2004) 'Novel patch-/matrix interactions: patch width influences matrix use by birds', *Oikos*, vol 107, pp634–644

United Nations (1992) The Convention on Biological Diversity, www.biodiv.org/convention/convention.shtml

UNFCCC (United Nations Framework Convention on Climate Change) (2005) Conference of the Parties, Eleventh Session, FCCC/CP/2005/L.2, available at http://cdm.unfccc.int/

Valkonen, S. and Ruuska, J. (2003) 'Effect of *Betula pendula* admixture on tree growth and branch diameter in young *Pinus sylvestris* stands in southern Finland', *Scandinavian Journal of Forest Research*, vol 18, no 5, pp416–426

Vandermeer, J. H. (1989) *The Ecology of Intercropping*, Cambridge University Press, Cambridge

Vilà, M., Vayreda, J., Gracia, C. and Ibáñez, J. J. (2003) 'Does tree diversity increase wood production in pine forests?' *Oecologia*, vol 135, pp299–303

Vilà, M., Vayreda, J., Comas, L., Ibáñez, J. J., Mata, T. and Obón B. (2007) 'Species richness and wood production: A positive association in Mediterranean forests', *Ecology Letters*, vol 10, pp241–250

Walker, B., Kinzig, A. and Langridge, J. (1999) 'Plant attribute diversity, resilience, and ecosystem function: The nature and significance of dominant and minor species', *Ecosystems*, vol 2, pp95–113

Watt, A. D., Stork, N. E. and Bolton, B. (2002) 'The diversity and abundance of ants in relation to forest disturbance and plantation establishment in southern Cameroon', *Journal of Applied Ecology*, vol 39, pp18–30

Wickens, G. E. (1991) 'Management issues for development of non-timber forest products', *Unasylva*, vol 165, p42

Wickens, G. E. (1994) 'Sustainable management for non-wood forest products in the tropics and subtropics' in FAO (ed.) *Readings in Sustainable Forest Management*, FAO Forestry Paper, 122, FAO, Rome, pp55–65

Williams, B. (1996) 'Total, organic and extractable-P in humus and soil beneath Sitka spruce planted in pure stands and in mixture with Scots pine', *Plant and Soil*, vol 182, pp177–183

Yachi, S. and Loreau, M. (1999) 'Biodiversity and ecosystem productivity in a fluctuating environment: The insurance hypothesis', *Proceedings of the National Academy of Sciences of the USA*, vol 96, pp57–64

Yi, H. and Moldenke, A. (2005) 'Response of grown-dwelling arthropods to different thinning intensities in young Dougas-fir forests of Western Oregon', *Environmental Entomology*, vol 34, pp1071–1080

Yuan, H. W., Ding, T. S. and Hsieh, H. I. (2005) 'Short-term responses of animal communities to thinning in a *Cryptomeria japonica* (Taxodiaceae) plantation in Taiwan', *Zoological Studies*, vol 44, pp393–402

6
Smallholder plantations in the tropics – local people between outgrower schemes and reforestation programmes

Benno Pokorny, Lisa Hoch and Julia Maturana

Introduction

While natural forests in the tropics continue to disappear at unabated speed, plantations gain increasingly in importance for providing environmental goods and services (Varmola and Carle, 2002; Evans and Turnbull, 2004; Homma, 2005; Bacha, 2006). Many countries in the tropics have set up afforestation and reforestation projects to improve degraded forest lands, fight desertification, prevent soil erosion and maintain water quality (Blay et al, 2004; MINAG and INRENA, 2005; Chokkalingam et al, 2006). As plantations promise high yields with simple technologies (Homma, 2005; Almeida et al, 2006), for decades governments and development organizations have been investing in exploring the potential of plantations for smallholders (UN, 2002; Chokkalingam et al, 2006). These efforts have included social programmes in large plantation companies, so-called outgrower schemes (Mayers and Vermeulen, 2002; Wightman et al, 2006) as well as the promotion of plantations managed by smallholders themselves.

So far, experiences from smallholder plantations have not fulfilled the high expectations. For example, many of the smallholder plantations established by social forestry projects since the late 1970s in tropical Asia and Africa to produce fuelwood and fodder no longer exist (Pandey and Ball, 1998). In the Philippines, plantations established a century ago still have not produced marketable products (Chokkalingam et al, 2006). Technically successful plantation programmes benefiting companies have affected local communities negatively by excluding people from former forest fallows, which were important for grazing livestock and as a source of wood and non-wood forest products (NWFP). Conflicts between plantation companies and displaced smallholders are commonplace in many areas of the world (Carrere, 1998; Mattoon, 1998; Eraker, 2000; Barr, 2001). Social conflict is expected to rise as plantation areas and human populations increase (WRM, 1999; Nawir et al, 2007).

In this chapter, we review the literature and empirical information from South America, Asia and Africa and present our field data collected in the Amazon region to contribute to a better understanding of the potential of plantations for smallholders living in the rural tropics. First, this chapter outlines common smallholder categories as a point of reference for analysis. Subsequently, plantations are classified by establishment scheme to structure the discussion of processes and consequences of plantation development perceived by smallholders. Then, the potential of plantations for smallholders is analysed focusing on programmes and projects typically promoted by national governments and international donors in cooperation with non-governmental organizations (NGOs), but also on outgrower schemes promoted by pulp companies (FAO, 1998). Based on these findings, the conditions for the success of smallholder plantations are discussed, and existing potentials compared to those of traditional tree growing schemes. We conclude that the significant advantages of locally developed low-input systems for growing trees should be considered more closely in external initiatives for local development.

Smallholders in the tropics: demands and capacities

In keeping with the general goal of poverty alleviation as stated at the world summit in Johannesburg in 2002 (UN, 2002), we use the term *smallholder* for people living in the rural tropics who own small areas of land that they cultivate for subsistence or commercial purposes, relying principally on family labour. In our simplified analysis, we distinguish between two idealized smallholder types: traditional communities and individual farmers.

The *traditional communities* described here have a long tradition of land use. In addition to smaller, individually managed properties, they often have common access to forest areas. In the Amazon, for example, the communities may own up to several thousand hectares of land (D'Antona et al, 2006). Often, these communities show an elevated level of social organization, which facilitates the common use of resources (Chibnik and de Jong, 1989). In many cases, the communities have developed diversified production systems incorporating both agricultural and forested land use. Extensive slash and burn systems combined with smaller, more intensively managed patches of less than 1ha are typical. The collection of forest products, in particular fire-wood and NWFP including hunting is important. Traditional communities tend to live in remote, still-forested landscapes with a low level of infrastructure. Far from urban centres and markets, livelihood strategies are based mainly on the use of accessible natural resources. They may also sell products at local markets. The flexibility to respond to changing environments or emergencies is an important feature of their production systems. Ecologically stable landscapes are fundamental for maintaining the productive potential of their land use systems. Families have little access to capital, and are widely excluded from markets and public services. Consequently, the families are extremely interested in cash income or employment opportunities. Furthermore, better access to quality community services, particularly health and education (UN, 2008), is essential. In particular, those communities located in remote areas require protection from external interest in their resources (RRI/ITTO, 2009).

The term *individual farmers* refers to farmers working individually in family units who focus on the cultivation of agricultural crops for local markets. They usually manage smaller plots than traditional communities. Often they have acquired land through initiatives promoting large settlements or individual efforts to improve their quality of life (Marquette, 2006). They are located in already fragmented landscapes with adequate infrastructure and connection to markets. Some families also have a complementary off-farm income. Forest use supplements agriculture. The economic status of families depends mainly on the markets for their agricultural products and, consequently, varies from year to year (Marquette, 2006). As they are more self-reliant, their social safety net is weaker than that in traditional communities, and depends on the degree of organization into cooperatives. In some cases, such organizations have managed to negotiate individual farmers' interests effectively in areas such as transport, credit and technical support. Farmers participate in the market economy and have access to community services such as schools and health care. However, they do not have capital available for larger investments. Like traditional communities, they are interested in cash income, better access to better quality services and they rely on ecologically stable landscapes for maintaining their production systems. An effective and flexible use of their relatively small properties is crucial. Due to the seasonality of agricultural production, additional income opportunities are important, especially in the rainy season.

Other important categories of rural populations, in particular landless families and indigenous groups are considered only indirectly. To a certain extent, statements about individual farmers may apply to landless farmers/families who are often involved in the cultivation of others' land, while lessons learnt about traditional communities to a certain extent may be valid for some indigenous groups.

Plantations and tree-related production systems of smallholders

According to FAO (1998), plantations are forest stands established through afforestation or reforestation by planting (or seeding) introduced species or one or two native species in intensively managed stands. In a broader sense plantations also include all the trees that people grow. To refine this definition in the context of smallholders, we classified plantation types according to establishment schemes in terms of establishing, maintaining and using the trees (Table 6.1). Typical features of these establishment schemes as well as the principal implications for smallholders are described below.

Table 6.1 *Classification of plantations and tree-related production systems*

Size	Species	Technologies	Production goals	Growers
ENTERPRISE-INITIATED				
Fast-growth fibre plantations				
Up to 10,000 hectares	Exotic fast-growing species[1] (more than 10m³ per ha per yr)	Sophisticated	Fibre (mainly pulp for paper and boards)	Companies, entrepreneurs
Timber, wood and non-wood forest products (NWFPs)				
Up to thousands of hectares	Domestic high-value timber species, firewood, NWFPs,[2] less exotics[1]	Medium	Timber, wood and fruits	Entrepreneurs, companies, estate
Outgrower schemes (mainly fast-growth and fruits)				
up to 20ha	Exotic fast-growing species or NWFPs	Medium	Fibre	Smallholders in cooperation with companies
DONOR (GOVERNMENT AND NON-GOVERNMENT)-INITIATED				
Reforestation programmes				
Up to 100,000 hectares	Preferential domestic species, varying in accordance to geographic conditions, plus some exotics	Low–medium	Protection of erosion, water (quality and floods) and biodiversity	Public institutions, smallholders
Agroforestry systems				
Up to 20ha	High-value timber species, firewood, NWFPs, as well as legumes and shadowing tree species in combination with agricultural crops	Medium–high	Timber, firewood, forage, rubber, fruits, rattan, bamboo, medicinal plants	Smallholders, medium farmers
Credit programmes for entrepreneurs				
Up to several hundred hectares	High-value timber species, firewood, NWFPs	Low–medium	Private income, raw material	Medium to big farmers
Less than 5ha	High-value timber species, firewood, NWFPs	Low–medium	Private income	Smallholders, medium farmers
Experimental plantations				
0.1ha up to a few hectares	Native species, agroforestry systems	Medium	Knowledge	Research institutions, smallholders

Table 6.1 *continued*

Size	Species	Technologies	Production goals	Growers
SMALLHOLDER-DRIVEN				
Production forests				
Up to 20ha	High-value timber species, firewood, NWFPs	Low	timber, firewood, rubber, fruits, rattan, bamboo, medicinal plants	Smallholders
Agroforestry systems				
Usually between 0.5 and 5ha	High-value timber species, fire-wood, NWFPs, as well as legumes and shadowing tree species in combination with perennial crops such as coffee, cacao etc.	Low	timber, firewood, forage, rubber, fruits, rattan, bamboo, medicinal plants	Smallholders
Homegardens				
0.1–10ha	Multi-use domestic tree species	Low	Multi-use for subsistence; seldom commercialization	Smallholders
Growing of single trees				
Single trees (few up to several thousand trees)	Multi-use domestic tree species	Low	Multi-use for subsistence, financial reserves	Smallholders

1 e.g. *Pinus radiata, Pinus patula, Pinus merkusii, Araucaria augustifolia, Copressus* spp.; *Eucalyptus* spp.; *Acacia mangium, Acacia* spp.

2 e.g. *Heva brasilliensis; Cryptomeria japonica; Chamaecyparis obtuse; Lavix leptolepis; Ginelina arborea, Paraserienthes falcataria*

Enterprise-initiated

Private investors, such as large pulp companies, sawmills or fruit companies, promote enterprise-initiated plantations, which are often subsidised because they are expected to create economic development opportunities for rural areas. For example, many existing pulp plantations established in Brazil between 1966 and 1988 received federal support (Fiscal Incentives for Forestation and Reforestation – PIFFR), and the plantation boom in countries such as Uruguay and Chile in the 1990s was encouraged by massive international funding. In addition to direct payments, subsidies included road construction, finance and reduced taxes, often by up to as much as 50 per cent (Cossalter and Pye-Smith, 2003; Nilsson, 2003; Bacha, 2006).

Generally, two types of plantations can be distinguished in this category: fibre plantations and plantations for timber and NTFPs.

- *Plantations established for the production of fibre for pulp* are initiated mostly by medium- to large-sized companies, including estate management companies (Cossalter and Pye-Smith, 2003). Rotation periods are short, and vary from 4 to 15 years. The plantations are managed with sophisticated technologies such as polytube systems, single-cell container systems, intermediate treatments and mechanized harvest operations. Most companies produce their own seedlings, sometimes up to 50 million per year (Maturana, 2005). The control and prevention of fire and disease is highly important and is handled by specialized departments.

- *Plantations for wood and NTFPs* are initiated by larger companies, individual entrepreneurs, as is the case for the coffee (*Coffea*) and cacao (*Theobroma cacao*) plantations in Africa, Latin America and Asia (Rice and Greenberg, 2000; Baffes et al, 2004), as well as estates, for example many of the rubber (*Hevea*) plantations in Asia (Ali et al, 1997). These plantations vary significantly in size (from small patches of less than 100ha up to thousands of hectares) and in the level of technology and organization depending on the product and socio-environmental conditions. Mostly, the technology used is less sophisticated than in pulp plantations.

For local people, employment opportunities are the most important benefit from this kind of plantation. However, such opportunities are often limited to unqualified labour during seasonal peaks, sometimes only for a few weeks each year. In comparison with other land uses the employment levels are relatively low and rarely exceed 1–3 direct and a further 1–3 indirect jobs for every 100ha planted (Cossalter and Pye-Smith, 2003; Schirmer and Tonts, 2003; WRM, 1999; Aracruz, 2001; WBCSD, 2001). In particular, Forest Stewardship Council (FSC)-certified enterprises provide good working conditions, often accompanied by the provision of education and health services. A few companies also run programmes to provide these services to local families living near the plantations. To simplify, the social impact of pulp plantations depends on two key factors: population density and former land uses. Both are influenced by a third parameter: the soil fertility (Figure 6.1).

Population density	Origin	
	Replacement of natural forests	Reforestation on degraded land
High	More negative	May be positive May be negative
Low	Indifferent	More positive

Figure 6.1 *Social effects of enterprise-initiated plantations*

If population density is high and plantations have replaced natural forests, local impacts tend to be more negative because, compared to original land uses, the plantations provide relatively few jobs, generate locally irrelevant products for export and tend to diminish the environmental quality of the landscape. Even deforested land provides essential goods and services to locals and tends to achieve a higher environmental value in the course of secondary succession. This is even more relevant as commercial investors generally avoid planting on strongly degraded soils for economic reasons. Furthermore, in populated areas, the probability of conflict over land tenure is always high. Indirect benefits to local people are confined to improved main roads, since investments in health and education are normally restricted to employees. If population density is low, the social balance of plantations tends to become more positive. But conflicts with traditional land-use practices and property rights remain probable, and plantations also limit the future options for local smallholders for a long time. Nevertheless, although the effects of plantations on local people may occasionally be dramatic, they are mostly indirect and often irrelevant. Any benefits are basically limited to the relatively low number of employees.

Outgrower schemes

First implemented in South Africa during the 1960s by pulp industries to guarantee the availability of raw material (see Box 6.1), outgrower schemes became important as a mechanism to soften existing conflicts with local communities about resource access, land tenure and the effects of environmental damage (Desmond and Race, 2000; Cyranoski, 2007). Outgrower schemes are production partnerships encouraging smallholders to grow trees on their own land as a source of raw material for industry. Generally, the company offers technical and financial support to establish and maintain the plantation, and ultimately purchases the products from the smallholder. In some cases the company also provides marketing and production services. The two most typical products of outgrower schemes are pulpwood and fruit. Outgrower schemes are usually established in areas with good infrastructure, clearly defined land tenure and favourable environmental conditions, and provide attractive income opportunities for the farmers (Mayers and Vermeulen, 2002; Nawir et al, 2002; Vidal and Donini, 2004)

Donor-initiated

In many regions of the world, international and national donors, in collaboration with governmental extension agencies and NGOs, support tree planting by donating tree seedlings, providing training courses, technical advice in the field and, in some cases, financial support, for example in the form of favourable loans or as payments for local working input. Donors commonly support reforestation programmes for environmental purposes as

---

Box 6.1 *Outgrower schemes*

Well-documented examples of outgrower schemes are the pulp companies Mondi and Sappi in South Africa. Here, around 10,000 smallholders grow *Eucalyptus* spp. on areas between 1.5 and 2.7ha in rotations of seven years. The companies provide seedlings, credit, fertilizer and technical assistance, and ultimately acquire the harvest for pulp production. The outgrowers receive payments before planting and after successfully completing each operation specified in the production programme, such as marking, ploughing, pitting, planting, fertilizing, weeding and fire protection. The money paid to the outgrower after completing each operation is essentially a loan advanced against the value of the final product (Mayers and Vermeulen, 2002). In the year 2002, outgrowers produced about 10 per cent of the two companies' mill consumption (Hall, 2003).

---

well as agroforestry projects with a social emphasis.

Reforestation programmes focus on environmental goals, such as the rehabilitation of degraded land to ensure water quality and prevent erosion. Yet production also plays a role. Such programmes are established mainly by governments and/or environmental NGO initiatives, or are legally enforced (Thacher et al, 1996; Sayer et al, 2004; MINAG and INRENA, 2006; Ministerio del Ambiente, 2006; Nawir et al, 2007). In the attempt to generate relevant outcomes, reforestation programmes tend to cover huge areas. Administration units are established temporarily to coordinate finances and logistics. Smallholders may become involved in the planting operations, and often in activities for fire control and other protection issues. To attract the interest of landowners, the incentives for participation are limited, often to the provision of plants and technical assistance free of change. Depending on the former land use and land rights, local people may receive monetary compensation (Almeida et al, 2006).

Agroforestry projects aim to produce both forest products and agricultural crops while maintaining soil fertility (Dubois, 1996; Pattanayak and Mercer, 2002; Puri and Nair, 2004). International donors, particularly in collaboration with NGOs, promote agroforestry projects as an opportunity to improve local livelihoods while providing environmental benefits (UNDCP, 1997; Vivan et al, 2002; Browder et al, 2005). Agroforestry systems can include a wide variety of tree species, including trees for the production of forage, fruits, rubber, timber or firewood as well as trees for the provision of shade or to support other plants, or those planted mainly for environmental purposes. Normally trees are planted at wide spacings to allow the cultivation of agricultural crops below. Trees may also be planted in lines on the edges of agricultural fields, as for example teak trees in Indonesia (Maturana, 2006). Some agroforestry techniques have a long tradition in the tropics (Wiersum, 1982; Nair, 1987). Although agroforestry systems are explicitly designed for smallholders, most of them are technically sophisticated (Box 6.2) and demand drastic changes in

---

Box 6.2 *The 'Götsch' agroforestry system*

The agroforestry approach developed by Ernst Götsch in subtropical north-eastern Brazil has been adapted widely by several development projects in South America. The principle of the Götsch approach is to make use of natural succession dynamics to achieve an abundant and diverse production while maintaining soil fertility without fertilizer and chemical use and without the need to fight disease or pests. Priority is given to a wide range of annual and perennial food crops and fruits adapted to the specific local ecological conditions. The tree component is less important. Typical crop species are rice, corn, beans, tomato, manioc, papaya, banana, cacao, coffee, citrus trees, palms and legumes as well as mahogany and other high-value timber species. All species are planted at high densities to respond effectively to all niches in the ecosystem. When the growth of one species declines another species take over. Weeding occurs selectively. Regular pruning adds more organic material to the system and regulates the light available for the development of different plant groups. The system is expected to contribute to the long-term stability of the ecosystem and the maintenance of biodiversity while also providing continuous opportunities for cash income. The Götsch system requires considerable expertise (Milz, 1998).

---

land-use practices from the farmers (Hoch, 2009).

Principally, initial and continuous support strategies can be distinguished (Hoch, 2009). A key element of initial support strategies is the distribution of tree seedlings produced in large nurseries. Initial support initiatives often provide technical guidelines, occasionally technical assistance and, in a few cases, financial incentives for planting. Support, however, is confined to the establishment phase. Most reforestation programmes fall in this category. Other organizations, often those involved in agroforestry projects, provide more continuous support through to the commercial phase. These continuous support initiatives normally manage to establish personal contacts with the families and often provide better-qualified assistance. Local demonstration plots are often set up and, in some cases, local people are engaged as extension agents. However, this more intensive support tends to concentrate on few, easily accessible families who often are economically better off.

Several governments and banks also support plantations with specific *credit programmes*. These initiatives usually have a commercial focus, and the funding agencies expect a financial return (Teixeira, 2005; Bacha, 2006). Nevertheless, in addition to offering favourable conditions to smallholders, they may also provide technical assistance. Frequently, these programmes are bureaucratic with relatively high transaction costs (Cacho et al, 2005). Consequently, this option is more appropriate for small entrepreneurs. Often donors combine rehabilitation and commercial goals in the expectation of creating win–win situations in so-called integrative projects (Box 6.3). Currently, the evolving CO_2 market generates significant additional resources for this type of initiative (Montagnini and Nair, 2004; Larsson et al, 2007).

> **Box 6.3** *Smallholder credits for reforestation in Vietnam*
>
> The cooperation between the Vietnamese government and the German financial organization (KFW, development bank) addresses the issue of deforestation and soil erosion in north and central Vietnam with an innovative afforestation model. The concept is based on participation, private sector initiative, quality monitoring and focuses on income generation for poorer households. Based on participatory land-use planning, interested farmers are provided with plots of a few hectares for afforestation. During the non-productive period before the first harvest, the farmers are supported in kind (seeds, equipment, etc.) as well as by training and extension services. In addition they receive direct payments according to their individual performance with the production plan. Therefore the donor establishes saving accounts for each farmer at the Bank for Agriculture and Rural Development, from which they receive payments in keeping with the results after regular external monitoring. The cooperation has implemented five afforestation programmes costing €38 million in total. In view of the first successful experiences, the reforestation of more than 100,000ha for the benefit of about 65,000 households is planned in future (Kuchelmeister and Huy, 2004).

Smallholder initiatives

Smallholders also grow trees on their own initiative in small tree plantations from less than one up to a few hectares (Smith et al, 1996; Rice and Greenberg, 2000), yet more frequently in agroforestry systems, homegardens or singly. Plantation products may contribute to household subsistence, or are sold on local markets. In comparison with the externally motivated establishment of plantations, the number of planted trees per hectare is generally much lower. Often smallholders use naturally regenerated trees (Peck, 1982; Nair, 1987; FAO, 1998; Kleinn, 2000; Pinedo-Vasquez et al, 2001; Hoch et al, 2009). The following paragraphs describe the main types of tree growing initiated by smallholders.

Production forests are typically established by smallholders able to invest some capital, and who have sufficient land and time available. They are also motivated by the existence of attractive markets for the tree products. Consequently, many initiatives tend to originate from local industries with a high demand for forest products, as for example the teak plantations in Java (Maturana, 2006) or occasionally from access to national or international markets generally encouraged by traders or NGOs. These plantings nearly always require low input and are technically poorly managed. Therefore productivities are low and growing periods relatively long. Normally the trees provide a minor complementary source of income, or are used for non-commercial purposes.

Agroforestry systems initiated by smallholders reveal a significantly lower level of complexity compared to those usually promoted by development organizations. In the absence of technical and scientific expertise. Smallholders pragmatically adapt traditional strategies, as for example slash and burn agriculture in the tropics (Hecht, 1982; Miller and Nair, 2006). Here, trees and forests are used mainly for soil improvement while growing forest products plays a minor role. Yet many indigenous and traditional communities have developed quite sophisticated permanent agroforestry systems (Padoch et al, 1985). Agroforestry systems combining perennial crops with trees are widely distributed (Box 6.4). In these systems, trees, while providing products for subsistence and markets, are planted mainly for the provision of shade (Mussak and Laarman, 1989, Ramirez et al, 1992; Neto et al, 2004).

Box 6.4 *Agroforestry systems initiated by smallholders*

Along the Trans-Amazonian highway in Brazil, where families in the late 1970s received plots of 100ha, some farmers have developed lucrative agroforestry systems. Apart from cultivating annual crops for their own consumption, farmers began cultivating perennials; mainly cacao and also coffee and pepper on a small scale. Guided by the Brazilian Research Institute for Cacao (Comissão Executiva de Planejamento da Lavoura Cacaueira – CEPLAC), farmers planted exotic leguminous trees in the cacao fields for the provision of shade. One farmer began experimenting with planting high-value timber trees such as *Swietenia macrophylla* King, *Cedrela* sp. and *Tabebuia* sp. Motivated by the excellent performance of these experiments and expecting attractive prices for high-value timber, other farmers also began to plant up to 100 trees per hectare in their cacao fields, applying different species composition and planting densities. Over time, the natural regeneration of high-value tree species became an important feature of local cacao production systems. Generally the preferred trees are expected to achieve target diameters in 20–50 years. Due to the success of the system, it was adapted by CEPLAC for more widespread use (Teixeira, 2005; Hoch, 2009).

Homegardens are intensively managed areas around houses consisting of trees cultivated together with annual and perennial agricultural crops and livestock. Homegardens are commonly found throughout the tropics. The average size of a typical homegarden is less than 0.5ha, with large numbers of tree and herbaceous species in a multi-strata arrangement. Although there is remarkable similarity in the type and nature of herbaceous crops among homegardens globally, the tree species change according to environmental and socio-cultural factors. The vast majority of homegardens are for subsistence. However, in more densely populated regions especially, and if local markets offer sufficiently attractive prices, as for example in the case of rubber in Java, smallholders use homegardens as a complementary source of income. Generally homegardens produce relatively low but continuous yields, and provide high flexibility at low risk (Fernandes and Nair, 1986; Kumar, 2006; Miller et al, 2006; Mohan et al, 2006).

Commonly smallholders grow *single trees* of species with domestic or commercial value on their land using a wide range of technologies (Alcorn, 1990; Budowski, 1990; Byron, 2001; Hoch et al, 2009) such as planting, transplanting and promoting or simply tolerating natural regeneration (Alcorn, 1990; Peck, 1990; Sears et al, 2007). In the Ecuadorian Amazon, for example, two-thirds of the trees found in coffee fields and in pasture originated from natural regeneration (Ramirez et al, 1992). While fruit trees generally play an important role, many farmers also consider slow-growing but valuable tropical hardwood species such as mahogany or tropical cedar (Browder et al, 1996; Smith et al, 1996; Pichon, 1997; Simmons et al, 2002; Almeida et al, 2006). The trees, normally grown with the agricultural crops, allow a relatively high agricultural production during the first years. Even after 3–5 years, the production of shade-tolerant permanent crops such as cacao and coffee can be maintained.

Plantations and smallholders

This section evaluates the potential of externally initiated characterized by a systematic, relatively dense planting of one or a few species by exploring two questions (Figure 6.2): first, to what extent do the benefits generated by plantations correspond to smallholders' demands, and second, do smallholders have the capacity to explore these potential benefits?

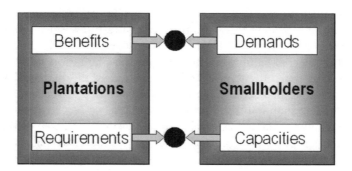

Figure 6.2 *Analytical framework to explore plantation potential for smallholders*

Potential benefits of plantations for smallholders

In view of the smallholders' demands outlined in the second section above, the following sections systematically analyse the relevance of externally promoted plantations in relation to the different benefit categories, in particular: (1) the generation of cash income urgently needed to participate in the market economy and to accumulate capital for investments; (2) the provision of environmental services as a basis for maintaining fertility and the natural capacity of smallholders' production systems; (3) the effectiveness of using scarce resources and the ability to respond flexibly to emergencies; (4) security in the sense of reduced risks; and, finally, (5) the potential indirect benefits such as improved access to education, health, credits, electricity, streets and markets.

Cash income

Without doubt, income is one of the main expectations from the generation of plantations for smallholders, particularly in view of growth rates of more than 20m³/ha/yr and export prices for high-value timber of up to US$500/m³ (Lentini et al, 2005). However, experiences show that, for a number of reasons, smallholders normally do not have real opportunities to access these benefits as they are rarely involved in value-adding processes and sell their products mainly as raw material for relatively low prices. Financial attractiveness also suffers from the relatively high initial investment in plants and fertilizer. Furthermore, technical errors as well as inappropriate, or lack of, silvicultural treatments impact negatively on the performance and growth rates of the trees. Most critical are threats from fire, however, pests and droughts and fire, which are very common in many tropical regions (Niskanen, 2000; Byron, 2001; Hoch, 2009), as well as marketing risks. In fact, at the time of establishment, the prices for plantation products are usually unpredictable due to long production periods and the absence of relevant markets. Income generation from perishable products such as fruits depends, in addition, on excellent logistics and an adequate infrastructure to ensure quick sales (Shanley et al, 2002; Shanley and Medina, 2005). As a consequence Hoch (2009), in their analysis of a selected sample of most promising plantation initiatives in the Amazon, found that only 1 in 100 farmers successfully managed to sell the products from their plantations and that, even in these few successful cases, the benefits were less than half than initially expected.

Clearly, the financial attractiveness of plantations increases with the extent to which the supporting organizations cover the necessary investments and guarantee attractive prices for the products. For example, in some of the few successful outgrower schemes, smallholders achieved average incomes of up to US$130/ha/yr (Mayers and Vermeulen, 2002). This is a significant source of income given that participating farmers manage more than 10ha. Also the growing of high-value timber trees promises high financial returns when the markets are adequate, as in the case of teak in Indonesia (Maturana, 2006). Some initiatives offer possibilities for reducing initial investments. For example, if the farmers produce their own seedlings, establishing costs may be reduced by up to 50 per cent. As a reduction in the significant cost components of transport and harvesting are unlikely, another common strategy for cost reduction is to sell the standing timber or wood (Hoch, 2009). However, while enhanced local participation ensures ownership and higher flexibility, it is often associated with the risk of technical deficiencies, which may significantly reduce the performance of the plantation, and, in the worst case, impact upon the success of the plantation growth.

In the case of credit programmes or payments, the most immediate benefit to the families is indeed the availability of funds. The funding received, however, is often invested in other land-uses or simply spend for consumption purposes. In the long run this may provoke serious conflicts when the credit dealers audit the application of credit funds. This situation, which is common, derives from the tendency for plantation programmes to focus exclusively on plantation issues, or rather from the lack of integrated approaches more strongly oriented to existing smallholder demands, which are often directed to other, more attractive, land-use options (Tamale et al, 1995).

Provision of environmental services
In particular in regions with little remaining forest cover and strongly degraded soils, plantations may provide important environmental services and contribute to ecosystem stability. In most cases, however, even highly degraded secondary forests may provide these services effectively, especially if compared to monoculture plantations with exotic species (Scherr et al, 2004). In addition, secondary succession runs naturally and thus requires little or no external input, which is an important advantage from the smallholders' point of view (Hoch et al, 2009).

The more degraded the soils are, and the slower the process of natural recovery, the more positive the effect of plantations on ecosystem stability. Nevertheless, from a financial point of view, it is generally not attractive to grow trees on degraded soils as growth rates are slow and natural risks high. Thus, smallholders, like large companies, prefer good soils for tree growing to avoid high fertilizer and pesticide inputs (Hoch et al, 2009). Consequently, the environmental contribution of plantations established for the generation of income is potentially low and may be even negative (see Chapter 5).

Efficiency and flexibility
For smallholders, the effective and flexible application of their scarce resources is fundamental. Plantations may provide an interesting opportunity to use existing resources more effectively to generate additional income, thereby improving smallholders' flexibility to respond to market opportunities. This is particularly true if markets for plantation products are well established and inexpensive transport is available throughout the year. Naturally, management options are most flexible when plantation management and sales are not regulated by laws or agreements with potential buyers. Thus, the need for formally approved management plans, production quota or specific requirements on quantities, qualities or timeframes significantly reduce attractiveness for smallholders over agriculture or (illegal) harvesting of natural forests.

However, plantations may also weaken the flexibility of local families, if they replace the diverse, low-input 'traditional' production components at a larger scale. A strong focus on the plantation product at the neglect of agricultural crops can make families more vulnerable to price changes. This may occur, in particular, when plantations are promoted strongly and subsidized by payments and credits, and receive intensive technical support (Hoch, 2009). In these cases, smallholders favour plantations even if they are non-competitive compared to traditional land uses. There are several examples where smallholders faced major difficulties in re-establishing their traditional land-use systems after the crash of plantation markets and, in extreme cases, even lost the basis of their livelihoods (Scherr et al, 2004). In this sense, the production of larger quantities of only one product, as in plantations, tend to bear other risks if compared to production of a wider array of products and services typical for traditional subsistence oriented production systems. Furthermore, from an ecological point of view, the cultivation of multiple products can be more meaningful, in particular in the tropics (Mohan et al, 2006). In the extreme, if forest products can be harvested at no cost in natural forests, investment in plantations for subsistence makes no sense at all from a smallholder's point of view.

Security
Tree growing may add long-term value to the land and thus contribute to the security of future generations (Fujisaka and White, 1998). In fact, it is often the expectation of legal recognition of traditional land and access rights that is the principal motivation for smallholders to join development projects and establish partnerships with environmental organizations (Medina et al, 2009). But growing trees, in particular for high-value timber, to a certain degree also increases insecurity as other actors may become interested in the valuable resources. The theft of wood is an issue. If planting is undertaken collectively, conflicts about sharing costs and benefits may arise. In extreme cases, the supporting organizations might grab a major part of the generated benefits, as, for example, in the cases of government agencies in India (Schmerbeck, 2003) or some pulp companies working with unfair agreements (WRM, 1999).

Doubtless, fire, traditionally used in many agricultural systems, is the single most important threat to plantations in the tropics. Fire is also used to clear land and as an instrument to enforce individual interests in land tenure conflicts or to overcome existing land-use restrictions (Goldammer, 2000). The problem is aggravated by poor fire management regimes frequently leading to uncontrolled accidental fires. Whereas the microclimate of primary forests make them relatively fire resistant, larger monocultures in particular suffer major devastation or even destruction from fires (FAO, 2001a). In Sumatra and Kalimantan, for example, more than 10 per cent of nearly 2 million ha of commercial plantations established since the mid 1980s were damaged by the 1997/1998 fire (FWI/GFW, 2002). For forest frontier areas in Latin America, a fire risk of 0.5–2.0 per cent per year is estimated (Simmons et al, 2002; Hoch,

2007). Also pests and diseases threaten plantations, in particular monocultures in the tropics, where climate and ecosystem dynamics favour calamities of insects and fungi. The risk increases with the size of the plantation, the rotation cycle and, most importantly for smallholders, the invested resources. In this sense, smallholder plantations fully financed by rehabilitation programmes, for example, reduce smallholders' sensitivity to risks, whereas financing of plantations through bank credits or repayable loans may cause immense problems to the families if the plantation is destroyed or damaged.

Pests and diseases also threaten plantations, in particular monocultures. The tropical climate, ecosystem dynamics and biodiversity favour calamities of insects and fungi. But, as forest plantations represent a relatively new land use in the tropics, there is still insufficient data available for a sound risk assessment (Nair, 2001; Cossalter and Pye-Smith, 2003). This lack of information means existing risks are unpredictable and plantations are an uncertain investment per se.

Finally, the social consequences of external interventions for the promotion of plantations should be considered. For smallholders, social networks are an important security feature as they provide basic livelihood services such as exchange of workload, division of tasks, support in the case of emergencies as well as employment opportunities and temporary provision of land to the landless poor. Although local social systems are relatively resilient, they are significantly affected by external interventions. In fact, the role of external organizations in the development of the framework of plantation, or other development projects provoke changes of culture, routines and traditions, which may strengthen or weaken the family social organization (Rogers, 2003). The actual effect depends mainly on the capacity of the communities to cope with change and challenges.

Indirect benefits

The collaboration with enterprises, NGOs and government organizations also has a number of indirect benefits that are highly relevant to smallholders (Medina et al, 2009). In view of the relatively strong isolation of smallholders from information, markets and public services, a ready contact to development and extension project agents may provide initial, or improved opportunities for locals to access relevant information and services (Medina et al, 2009) as well as to identify more attractive markets for their agricultural as well as plantation production (Scherr et al, 2004). The latter is particularly relevant if the plantation products are sold at local markets. In this sense, plantation projects may help to establish and strengthen local networks for the production, processing and consumption of local production as, for example, in the known cases of firewood in Madagascar (Raik and Decker, 2007), local timber markets in the Amazon (Pinedo-Vasquez et al, 2001) and fruits in China (Weyerhaeuser et al, 2006).

In particular, meeting other families and communities in the framework of plantation projects tends to stimulate the exchange of experiences and opinions, and may generate opportunities for becoming better organized and more actively involved in public policies. In particular, NGOs involved in plantation initiatives often promote the establishment of local organizations as mechanisms for negotiation and coordination. The existence of such associations also facilitates access to new credit programmes and other development initiatives and as well as acknowledgement of traditional tenure and access rights (Satterthwaite and Sauter, 2008)

In the case of outgrower schemes, smallholders may additionally benefit from company investments in the infrastructure required to ensure access to the new sources of raw material. To guarantee the success of the plantations promoted, the companies normally provide intensive training and technical support to the smallholders, which could also be useful for other land-use activities. In donor- or government-driven plantation programmes, this outcome is less likely as technical support is often restricted to the establishment phase.

Requirements vs. capacities

As shown above, plantations may offer interesting benefits to smallholders. Yet to attain these benefits, smallholders need to invest time and resources and require specific capacities. Thus, the possibility to explore the existing potential of plantations depends on the availability of the inputs and capacities required. This section critically assesses the extent to which smallholder resources and capacities correspond to those needed to develop plantations, particularly regarding finance, land, time, technical and managerial skills, and also analyses opportunities for balancing any discrepancies found.

Plantation requirements

Generally, plantations and agroforestry systems promoted by enterprises and development organizations implicitly require significant investment. The specific costs depend on a wide variety of parameters such as plantation size, planting density, management intensity and transport distances. Table 6.2 shows cost examples for smallholder plantations in China and Australia and provides comparative data about smallholder-driven tree-growing initiatives from a case study at the Amazonian highway.

Table 6.2 shows that the establishment of a plantation requires a significant investment. These costs vary strongly in relation to the type of nursery, transport distances, soil quality (need for fertilizer) as well as the tree species grown. Generally, seedlings from exotic tree species are less expensive than those from domestic ones as they are produced in smaller quantities often with more basic, less effective technologies. Along the Trans-Amazonian highway, for example, mahogany seedlings were sold for US$1.40 per seedling whereas 100 eucalypt seedlings cost less than US$10 in 2009 (Hoch, 2009). The cost of site preparation and planting in regular planting schemes is directly proportional to the size of the area planted. The application of fertilizer in fast-growing plantations is mandatory, even on fertile soils. For optimal growth,

Table 6.2 *Expenditures for plantations (excluding land costs)*

Activity	Commercial plantations			Tree growing	
	Activities	China[a] US$·ha⁻¹	Australia[b]	Amazonian highway[c] Activities	US$·ha⁻¹
Establishment	Plants, site preparation, planting, 3 fertilizer applications and weed control	619	1346	own workforce plants bought and planting	65 200
Maintenance	Fertilizing, weed control and thinning	319	226	included in agricultural treatment	0
Harvesting	Cutting and skidding	1000	not available	cutting and skidding	300
Transport	Up to 50km	5000	not available	up to 50km	5000

[a] *Eucalyptus* spp., 7 years rotation; data adapted from Xu (2003)
[b] *Eucalyptus* spp., 30 years rotation; data adapted from Heathcote (2002)
[c] 100 trees per ha, 20–50 years rotation, data adapted from Hoch, 2009

plantations require intensive treatments such as pest management, regular fertilization, thinning and pruning, which over the years also involve major costs. Finally, harvesting and especially transport are also costly. As a consequence, the total costs for establishing, maintaining and harvesting a fast-growing plantation may easily reach US$2000 per ha.

The management of forest plantations is time consuming, especially during the establishment phase and at the end of the production cycle. Compared to large companies, smallholders' working input per tree is even higher as they work without machinery, and opportunities for the specialization and division of work are limited. Generally, the work input per tree increases with smaller areas, lower mechanization and the need for individual treatments. Hoch, 2009 estimate, from their study in the Amazon region, that smallholders need about 60 days to establish one hectare of plantation for the production of timber and/or cacao, which involves site preparation, producing and planting 500 seedlings as well as weeding in the first year. Just watering the seedlings in the nursery requires more than one hour per day over several months. The same study concluded that, in view of this time input, farmers are able to establish a maximum of up to 2ha of plantation per year with the family workforce. Larger areas require (costly) external labour. During the two years subsequent to the establishment of the trees, weeding sometimes needs to be conducted four times to ensure the survival and growth of plants, which easily may aggregate to another 20–28 working days per hectare. Once the trees are tall enough to withstand the competition from secondary vegetation, the time input reduces significantly for most plantation types. However, for the production of high-value timber, for example, continuous silvicultural treatments such as thinning and pruning are necessary. The work input for harvesting depends on the product, the cutting diameter and degree of mechanization. Generally, harvesting of NWFP is more time intensive. In cupuazú plantations in Bolivia, for example, around 2–3 hours per day and ha are required to harvest the fruits over a period of 3–4 months a year (Hoch, 2009).

As mentioned above, plantations need to be established on fertile soils and on relatively large areas to ensure adequate productivity and a reasonable cost–benefit ratio. This is particularly true for products with relatively low unit values such as pulpwood. Planting on less fertile soils leads to sub-optimal growing conditions and demands investment in amelioration and fertilization, negatively affecting the financial attractiveness of plantations (Hoch et al, 2009).

In addition, specific technical skills are needed to manage plantations adequately, including activities such as the selection of seeds and plants, the treatment of seeds and seedlings, managing nurseries, thinning and pruning and harvesting. The management of plantations is also an organizational challenge and requires long-term planning and the proper organization of work. The treatment of seedlings in the nurseries, planting, the application of fertilizer, pest control and silvicultural treatments need to be carried out properly in clearly defined timespans. Also critical is the challenge of negotiating with external agencies for setting up the terms of cooperation and marketing is most critical (Scherr et al, 2004). In the case of outgrower schemes, the terms of cooperation are often pre-defined by the company (WRM, 1999; WBCSD, 2001). Agreements with traders and companies also may include the need to achieve certain quality standards as well as the delivery of certain quantities in pre-defined timeframes (Pokorny and Phillip, 2008). Often, the participation in plantation programmes, particularly if related to the transfer of resources, requires that families establish associations, which signifies another organizational challenge.

Smallholders' capacity to respond to plantation challenges

Many of the requirements listed above potentially exceed the financial, technical and human capacities of smallholders. In view of the notorious lack of readily available capital of poor families in the rural tropics (Byron, 2001; Sunderlin et al, 2005), the financial requirements represent the major limitation. Exceptions may be found within settler families with successful enterprise backgrounds, or those with capital from recent land sales. However, these families often act as small entrepreneurs systematically investing in the expansion of production areas and employing an increasing number of locals from outside their families (Godar, 2009).

The time requirements may also be unsuited to smallholders' capacities, since smallholder families have a full annual work calendar indicated by the fact that children are often heavily involved in production activities. The work pattern is determined by rhythm of nature with frequent work peaks when planting and harvesting agricultural crops. Nevertheless, there are also less intense working periods, normally during the rainy season, and also work-free periods as most work is carried out in the early morning and the late afternoon when temperatures are lower. There are also less intense working periods, normally during the rainy season. Furthermore, families with many adult children may have manpower available for extra activities. Hence, many smallholder families have some – however limited – time available during specific periods of the year. If sufficiently attractive, they also are able to adapt their working patterns (Ellis, 2000). Nevertheless, time requirements of plantations – especially if established at a larger scale – potentially conflict with the other production activities. Therefore, the more flexible the work input is, the higher the compatibility with existing capacities.

Regarding land requirements, the situation varies considerably for land requirements. Most smallholders, with the exception of some traditional and indigenous communities, by definition have limited land area available and only restricted access to resources outside their small individually owned patches of land. Consequently, the availability of land to grow trees is also restricted, and even more so if the existing natural forests are to remain untouched. As plantations grow best on fertile soils, they potentially compete with agricultural crops in smallholder farming systems. Even in the Amazon region, where many settlers own areas up to 100ha, the establishment of plantations at least involves the replacement of secondary forests that are a part of shifting cultivation schemes. Thus, the availability of larger areas of fertile land for smallholders is limited to the rare situations where traditional or indigenous communities have access to collectively owned, already deforested but only slightly degraded land.

Many authors have acknowledged the role of local farmers as experimenters (Barlett, 1980; Friedrich et al, 1994, Pinedo-Vasquez et al, 2001; Rogers, 2003; Sears et al, 2007), and doubtlessly many smallholders have accumulated considerable knowledge about the management of their resources, continually enhancing their knowledge through observation and experimentation so that their production systems gradually improve (Harwood, 1979; Barlett, 1980). Yet the technical requirements for plantations currently promoted by governments, donors and enterprises necessitate very different technical capacities and involve drastic changes to traditional natural resource management systems. This strongly contradicts local innovation patterns, normally a slow process of adopting and adapting new approaches depending on opportunities for experimenting and long-term observations (Rogers, 2003). In fact, most smallholder livelihood strategies are determined by annual routines based on the fulfilment of short-term objectives and spontaneous action in dealing with unpredicted events, such as fire, drought, flood, drastic changes in market prices, accidents, illness, etc. Although strategic planning for education, settlement activities and land-use options may occur, day-to-day practice still follows the annual working pattern and emphasizes the maintenance of flexibility required in view of the notorious lack of financial resources as well as the unpredictability of the future in a highly dynamic socio-economic context.

Finally, plantations imply the need for negotiations with more powerful actors engaged in the establishment and financing of plantations as well as the marketing of products. These negotiations are problematic for most smallholders (Sunderlin et al, 2005). External organizations, due to their experience and qualifications, consciously decide to work with plantations and collaborate with smallholders. Consequently, they have clearly defined and perceptions about the terms and conditions for cooperation. Companies and promoters of environmental projects and programmes are also highly qualified and usually have ample time to assess the implications and results of a specific initiative. In contrast, smallholders normally do not have access to relevant information and have fewer opportunities to use and interpret them. Furthermore they have no clear strategy about how to satisfy their demands and how to deal with plantations. As a consequence, they tend to accept the statements and promises made by external organizations without considering the future implications. This behaviour is often promoted by historically evolved paternalistic structures (Medina et al, 2009). In addition, smallholders often lack an effective organization for negotiating with these external organizations. Thus, individual families have neither the voice nor the power to effectively articulate their interests. As a consequence, issues such as prices, conditions for payments and risks usually are determined by the companies (Medina et al, 2007). Even with regards to indirect negotiations, as for example through market mechanisms, the intermediaries and buyers working within these monopolies or oligopolies tend to dictate the conditions of trade (Scherr et al, 2004). The comparative disadvantages of smallholders in negotiating with external actors often result in agreements that reflect the external interests more than those of smallholders. Serious conflicts may arise, if smallholders are not interested in, or simply unable to fulfil their negotiated duties and responsibilities when external actors enforce them. Once contractual agreements have been fulfilled, smallholders often discontinue the plantation initiative, in particular if the external support or control ends because it is unattractive or unfeasible (Pokorny and Johnson, 2008; Hoch, 2009).

Final considerations

Plantations may provide important benefits to smallholders, in particular they may generate cash income, increase the value of smallholders' land clarify tenure questions more quickly, improve access to credit and technical support, and contribute to ecological stability. Yet, to successfully deal with plantations, smallholders have to invest significant resources and have specific capacities available. These requirements normally exceed smallholders' technical, financial and managerial capacities not necessarily available. Consequently, the establishment, maintenance, correct management as well as the marketing of plantations is challenging and risky from a smallholder's perspective, and requires continuous external support. This, however, may be detrimental to traditional structures and management schemes imply the danger of getting dependent on external support. In this sense, plantations may contribute to a more effective use of scarce resources but also many block more attractive alternative land-use options. In the following section we analyse the attractiveness of plantations from the smallholders' perspective and define the key factors for success. In view of the limited feasibility of plantations for smallholders, we also highlight the potential of traditional tree-growing schemes.

Attractiveness of plantations for smallholders

Apart from cultural preferences, traditions and other 'non-rational' criteria, financial attractiveness is the key factor for smallholder decisions about land use. Yet for the generation of cash income, plantations show significant competitive disadvantages compared to agriculture. Even if supported by external organizations, plantations still require significant inputs of land and time. They represent long-term investments with the accumulation of risk over several years. Smallholders also need to wait a relatively long time for cash returns, which, due to insecure market situations, are often unpredictable. In contrast, although the cultivation of agricultural crops does not necessarily generate a higher financial return per hectare or per work input, it promises a more immediate income, allows flexibility to respond to changing markets, is based on existing traditional knowledge and skills and doesn't require long-term investment. Thus, in many situations agricultural land uses provide benefits more quickly, more flexibly and with a lower level of risk (Hoch, 2009).

Furthermore plantations consisting of only one, or a few – often exotic – tree species may provide fewer environmental benefits for smallholders than alternatives such as forest fallows, secondary forests or degraded primary forests. In still forested landscapes, for example, the regeneration of deforested land occurs relatively quickly (Nepstad et al, 1991). In contrast to plantations, these succession processes run naturally, and thus require little management input. Generally, secondary forests are also more diverse and ecologically more adapted to local ecosystems. Summing up, secondary forests, compared to plantations, have the potential to deliver environmental services of a higher quality at significantly lower costs.

Conditions for success

In view of possible disadvantages of plantations compared to alternative land uses the circumstances in which plantations present a viable economic option for smallholders should be carefully assessed. Doubtless, plantations are more attractive in landscapes already deforested with good infrastructure and clearly defined land tenure. It is also crucial that market risks are low; that is, that stable attractive markets already exist when the plantation is established. Also the opportunity to easily access local markets positively influences the viability of plantations for smallholders. Of fundamental importance is also the chance to establish fair, long-term partnerships with local actors involved in adding value to plantation products. In this respect, the existence of strong smallholder organizations is a key issue to ensure effective and favourable negotiations with relevant external players. The absence of powerful actors with competing land and resource management interests, as well as the existence of mechanisms that protect smallholders' interests are also helpful. Finally, limited profits from, or environmenal or legal restrictions to, competing land uses, especially agriculture, also enhance the viability of plantations.

In general terms, plantations tend to be more attractive for individual farmers than for communities, as farmers have stronger linkages to markets, are better connected to existing infrastructure and, furthermore, are more experienced in adopting and adapting technologies promoted by external actors. For smallholders living in remoter areas the immense market distances signify a critical barrier to viability of plantations at the outset. Yet, individual farmers also may have insufficient capacities, in particular of financial resources, land and organizational skills. Where the above-mentioned limitations to viable plantations occur, long-term technical assistance and adequate financing mechanisms are indispensable for success (Varmola and Carle, 2002; Almeida et al, 2006).

The potential of traditional tree-growing schemes

Growing trees is already an intrinsic part of most smallholder production systems (Smith et al, 1996; Simmons et al, 2002; Summers et al, 2004; Hoch et al, 2009), which shows that smallholders already have the interest, motivation and the skills and resources to grow trees for their own purposes. However, the smallholder-driven tree-growing initiatives generally show some important differences to the plantation schemes envisaged by external actors. Most important, nearly all smallholder-driven tree-growing schemes are low-input systems (Hoch, 2009) in which, instead of systematically planting seedlings of one species in a larger area, a variety of different techniques, ranging from natural regeneration through protection, transplanting or simply promotion of the species present, are adopted. Smallholder systems also avoid the application of fertilizer and pesticides, and also waive time-consuming treatments. In addition, smallholders confine tree-growing activities to periods when the agricultural workload is low, or combine tree-growing efforts with agricultural activities, for example weeding.

Due to the lower number of trees planted, local tree-growing schemes are naturally less expensive than the establishment of larger plantation areas. The costs per tree may also be lower due to the less sophisticated and more extensive nature of the technologies adopted. Especially when trees are grown together with perennial crops, the cost and time input is almost zero as the trees benefit indirectly from the regular management of the perennial crops (Hoch, 2009). These low-input strategies, which require minimal investment, apart from being applicable without external subsidies, are associated with a low level of risk and are extremely robust against price variations (Byron,

2001; FAO, 2001b; Shanley et al, 2002; Medina, 2004). On the other hand, local farmers often use less suitable genetic material and tend to completely ignore meaningful silvicultural treatments. This may negatively affect the performance of the plantation and, in the worst case, endanger the success of the growing activity (Almeida et al, 2006).

But smallholders do not necessarily grow trees for generating income. More commonly they grow trees for the production of fruits, medicine, timber and other subsistence purposes. Often the trees are simply favoured to provide environmental services to the agricultural crops such as for soil protection, windbreak, shade and nutrient enrichment (Summers et al, 2004). These benefits are especially important for smallholders with only little land (Nair, 1993). From a financial point of view, smallholders tend to prefer the production of NWFPs, as they provide more immediate and regular benefits than wood and timber. This is often more important than the prospect of a high total income in the long term. Also the cultivation of perennials such as *cacao* and – more recently – oil palm (*Elaeis*) tend to be highly lucrative (Hoch, 2009). Although generally avoiding high investments, when attractive marketing opportunities emerge, farmers have also proven their capacity to intensify, adapt and extend their traditional tree-growing systems (Hoch et al, 2008) as, for example, shown by the management of natural stands of açaí palms (*Euterpe oleracea*) in Brazil, and ita palms (*Mauritia flexuosa* L.) in several parts of the Amazon, camu camu (*Myrciaria dubia* (H.B.K.) McVaugh) in Peru, and teak (*Tectona grandis*) in Indonesia (Maturana, 2006).

For smallholders, growing single trees is better suited to their capacities than the management of plantations, as it excludes possible risks, combines well with daily routines and the flexible application of their own workforce, and depends on local technologies under their own control. Hence, for subsistence, forest goods and services as well as for marketing relatively small quantities, smallholders' traditional production systems tend to be much more effective and flexible than plantation schemes promoted by external organizations (Hoch, 2009).

Conclusions

In many tropical contexts, plantations are becoming more and more attractive as a result of ongoing deforestation and expanding infrastructure. However, their potential for smallholders is limited. Even in the most favourable conditions, the vast majority of smallholders would adopt forest plantations only to complement agricultural land-use options, which generally afford greater flexibility and lower risks. In most cases, the growing of single trees or groups of trees on their farmland corresponds better to smallholders' demands and capacities than the establishment of larger plantation stands. Although often poorly managed from a silvicultural point of view, growing single trees with minimal input is attractive to smallholders because it leaves them in control, avoids risks, ensures flexibility and – in the case of market access – also generates income.

Unfortunately, current efforts to promote smallholder plantations widely disregard the opportunity to support and optimize these traditional tree-growing practices. Instead, most externally driven initiatives deal with the diffusion of technology packages, which do not correspond to local realities, working rhythms and skills, and vastly ignore existing local knowledge (Pokorny and Johnson, 2008; Hoch, 2009; Nawir et al, 2007). Most of these observations are also valid for the diffusion of agroforestry systems defined by external experts. Although adapted to local conditions, and promising high flexibility with a large range of products, the effective use of resources and lower risks, most agroforestry systems promoted externally do not consider local capacities and interests, and generally ignore unfavourable conditions, in particular, the lack of attractive markets for the generated products.

In view of the existing incompatibilities, major financial, technical and managerial support is necessary to successfully diffuse the technical-institutional packages for plantations and agroforestry systems to smallholders. This, however, is rarely achieved by government and non-governmental organizations. Only companies offering outgrower schemes show high adoption rates because they have a vested interest in their success and the capacity to provide the required support. Nevertheless, in view of the conditions of collaboration generally determined by the companies, outgrower schemes might also be interpreted as companies simply renting smallholder areas and workforce for growing their trees (Desmond and Race, 2000).

Considering the many technical, economic and social pitfalls of plantations, it is worthwhile thinking more systematically about possibilities for promoting and optimizing local tree-growing systems. Yet such an approach would require a paradigms shift including the need to respect individual choices, to build on local technologies rather than transferring externally defined technological packages, as well as to switch from short-term development projects to more extensive but continuous supporting schemes under local control. Smallholders need time and space to carefully assess existing options as a basis for critical decision-making and the development of their own concepts and critical decision-making. Experiences show that smallholders then will decide on their own about if and how to intensify tree growing activities be it for the generation of income, for subsistence to provide environmental services as an asset for the future generation, or simply for aesthetic reasons.

References

Alcorn, J. B. (1990) 'Indigenous agroforestry strategies meeting farmers' needs', in A. B. Anderson (ed.) *Alternatives to Deforestation: Steps Toward Sustainable Use of the Amazonia Rain Forest*', Columbia University Press, New York

Ali, R., Choudhry, Y. A. and Lister, W. (1997) *Sri Lanka's Rubber Industry: Succeeding in the Global Market*, World Bank, Washington, DC

Almeida, E., Sabogal, C. and Brienza, S. (2006) *Recuperação de Áreas Alteradas na Amazônia Brasileira: Experiências locais, lições aprendidas e implicações para políticas públicas*, CIFOR, Belém

Bacha, C. J. C. (2006) 'The evolution of reforestation in Brazil', *Oxford Development Studies*, vol 34, pp243–263

Baffes, J., Lewin, B. and Varangis, P. (2004) 'Coffee: Market setting and policies', in M. A. Aksoy and J. C. Beghin (eds) *Global Agricultural Trade and Developing Countries*, World Bank, Washington, DC

Barlett, P. F. (1980) 'Adaptive strategies in peasant agricultural production', *Annual Review of Anthropology*, vol 9, pp545–573

Barr, C. (2001) *Political Economy of Fibre and Finance in Indonesia's Pulp and Paper Industries – Banking on Sustainability: Structural Adjustment and Forestry Reform in Post-Suharto Indonesia*, CIFOR and WWF's Macroeconomics Program Office, Washington, DC

Blay, D., Bonkoungou, E., Chamshama, S. A. O. and Chikamai, B. (2004) *Rehabilitation of Degraded Lands in Sub-Saharan Africa: Lessons Learned from Selected Case Studies*, Forestry Research Network for Sub-Saharan Africa, Accra, Ghana

Browder, J. O., Trondoli Matricardi, E. A. and Abdala, W. S. (1996) 'Is sustainable tropical timber production financially viable? A comparative analysis of mahogany silviculture among small farmers in the Brazilian Amazonia', *Ecological Economics*, vol 16, pp147–159

Browder, J. O., Pedlowski, M. and Summers, P. M. (2004) 'Land use patterns in the Brazilian Amazon: Comparative farm-level evidence from Rondônia', *Human Ecology*, vol 32, pp197–224

Browder, J., Wynne, R. and Pedlowski, M. (2005) 'Agroforestry diffusion and secondary forest regeneration in the Brazilian Amazon: Further findings from the Rondonia Agroforestry Pilot Project (1992–2002)', *Agroforestry Systems*, vol 65, pp99–111

Budowski, G. (1990) 'Homegardens in tropical America: A review', in K. Landauer (ed.) *Tropical Home Gardens: Selected Papers from an International Workshop held at the Institute of Ecology, Padjadjaran University, Bandung, Indonesia, 2–9 December 1985*, United Nations University Press, Tokyo

Byron, R. N. (2001) 'Keys to smallholder forestry in developing countries in the tropics', in S. R. Harrison and J. L. Herbohn (eds) *Sustainable Farm Forestry in the Tropics: Social and Economic Analysis and Policy*, Edward Elgar, Cheltenham

Cacho, O. J., Marshall, G. R. and Milne, M. (2005) 'Transaction and abatement costs of carbon-sink projects in developing countries', *Environment and Development Economics*, vol 10, pp597–614

Carrere, R. (1998) *Ten Replies to Ten Lies*, World Rainforest Movement (network), Montevideo

Chibnik, M. and de Jong, W. (1989) 'Agricultural labour organization in Ribereño communities of the Peruvian Amazon', *Ethnology*, vol 28, pp75–95

Chokkalingam, U., Carandang, A. P., Pulhin, J. M., Lasco, R. D., Peras, R. J. J. and Toma, T. (2006) *One Century of Forest Rehabilitation in the Philippines: Approaches, Outcomes and Lessons*, CIFOR, Bogor

Cossalter, C. and Pye-Smith, C.. (2003) 'Fast-wood forestry: myths and realities', *Forest Perspectives*, CIFOR, Jakarta

Cyranoski, D. (2007) 'Logging: The new conservation', *Nature News Feature*, no 446

D'Antona, A. O., Vanwey, L. K. and Hayashi, C. M. (2006) 'Property size and land cover change in the Brazilian Amazonia', *Population & Environment*, vol 27, pp373–396

De 'Nadai, A., Overbeek, W. and Soares, L. A. (2005) *Promises of Jobs and Destruction of Work. The Case of Aracruz Celulose in Brazil: Eucalyptus Plantations and Paper Mills/Production of Pulp Cellulose*, World Rainforest

Movement, Montevideo

Desmond, H. and Race, D. (2000) *Global Survey and Analytical Framework for Forestry Out-Grower Arrangements*, FAO, Rome

Dubois, J. (1996) 'The role of agroforestry in the sustainable development of the Amazon', Third Meeting of the Participants of the Pilot Program to Conserve the Brazilian Rain Forest (PP-G7), Bonn

Ellis, F. (2000) *Rural Livelihoods and Diversity in Developing Countries*, Oxford University Press, Oxford

Eraker, H. (2000) *CO2lonialism: Norwegian Tree Plantations, Carbon Credits and Land Conflicts in Uganda*, NorWatch, Oslo.

Evans, J. and Turnbull, J. (2004) *Plantation Forestry in the Tropics: The Role, Silviculture, and Use of Planted Forests for Industrial, Social, Environmental, and Agroforestry Purposes*, Oxford University Press, Oxford

FAO (1998) *FRA 2000: Terms and Definitions*, Working Paper 1, FAO, Rome

FAO (2001a) *Global Forest Fire Assessment: Forest Resource Assessment,* Working Paper 55, FAO, Rome

FAO (2001b) *Markets for High-Value Tropical Hardwoods in Europe*, University of Wales, Bangor, UK

Fernandes, E. C. M. and Nair, P. K. R. (1986) 'An evaluation of the structure and function of tropical homegarden', *Agricultural Systems*, vol 21, pp279–310

Friedrich, K., Gohl, B., Singogo, L. and Norman, D. (1994) *Farming Systems Development, a Participatory Approach to Helping Small-Scale Farmers*, FAO, Rome

Fujisaka, S. and White, D. (1998) 'Pasture or permanent crops after slash-and-burn cultivation? Land-use choice in three Amazon colonies', *Agroforestry Systems*, vol 42, pp45–59

FWI/GFW (Forest Watch Indonesia/Global Forest Watch) (2002) *The State of the Forest*, Forest Watch Indonesia/Global Forest Watch, Washington

Goldammer, J. G. (2000) *Forest Resources Assessment*, Working Paper 55, Forestry Department, FAO, Rome

Godar, J. (2009) 'The environmental and human dimensions of frontier expansion at the Transazon highway colonization area', PhD thesis, University of Leon, Leon

Granda, P. (2006) 'Monocultivos de árboles en Ecuador', www.wrm.org.uy/paises/Ecuador/Libro2.pdf (last accessed 21 May 2010)

Hall, C. (2003) 'Aracruz Celulose', panel on business and community, World Bank Forest Investment Forum

Harwood, R. (1979) *Small Farm Development: Understanding and improving farming systems in the humid tropics*, Westview Press, Boulder, Co

Hecht, S. B. (1982) 'Los sistemas agroforestales en la cuenca Amazónica: Práctica, teoría y límites de un uso promisorio de la tierra', in S. B. Hecht (ed.) *Amazonia. Investigación sobre agricultura y uso de tierras*, Centro Internacional de Agricultura Tropical, Cali, Colombia

Hoch, L. (2007) 'The dilemma of planting trees in fire-prone Amazon', www.waldbau.uni-freiburg.de/forlive/BoletinIV_home.html (last accessed 21 May 2010)

Hoch, L. (2009) 'Do smallholders in the Amazon benefit from treegrowing?' PhD thesis, University of Freiburg, Freiburg

Hoch, L., Pokorny, B. and Medina, G. (2008) 'Plantaciones forestales por productores familiares en la Amazonía', Policy brief, University of Freiburg, Germany

Hoch, L., Pokorny, B. and de Jong, W. (2009) 'How successful is tree growing for smallholders in the Amazon?', *International Forestry Review*, vol 11, pp299–310

Hoch, L., Pokorny, B. and de Jong, W. (forthcoming a) 'Financial attractiveness of tree growing for smallholders in the Amazon'

Hoch, L., Pokorny, B. and de Jong, W. (forthcoming b) 'Tree growing innovations in the Amazon: The challenge of adapting extension strategies to smallholders' decision-making processes'

Homma, A. K. O. (2005) 'Amazônia: como aproveitar os benefícios da destruição', *Estudos Avançados*, vol 19, pp115–135

Kleinn, C. (2000) 'On large-area inventory and assessment of trees outside forests', *Unasylva*, vol 51, pp3–10

Kuchelmeister, G. and Huy, L. Q. (2004) *Vietnamese–German Financial Cooperation Smallholder Forestry Project: Training Manual*, Margraf Publishers GmbH, Weikersheim

Kumar, B. M. H. (2006) *Tropical Homegarden: A Time-tested Example of Sustainable Agroforestry*, Springer, Dordrecht

Larsson, T. B., Barbati A., Bauhus J., Brusselen J. V., Lindner, M., Marchetti, M., Petriccione, B. and Petersson, H. (2007) 'The role of forests in carbon cycles, sequestration, and storage', IUFRO Newsletter

Lentini, M., Pereira, D., Celentano, D. and Pereira, P. (2005) *Fatos Florestais da Amazônia 2005*, Imazon, Belém

Marquette, C. M. (2006) 'Settler welfare on tropical forest frontiers in Latin America', *Population & Environment*, vol 27, pp397–444

Mattoon, A. T. (1998) 'Paper forests', *World Watch Magazine*, vol 11, no 2, pp20–28

Maturana, J. (2005) *Economic Costs and Benefits of Allocating Forest Land for Industrial Tree Plantation Development in Indonesia*, CIFOR Working Paper No. 30, Bogor

Maturana, J. (2006) 'Rethinking plantation forestry: Teak in Java', *CIFOR News*, vol 40, CIFOR, Bogor, Indonesia

Mayers, J. and Vermeulen, S. (2002) *Company–Community Forestry Partnerships: From Raw Deals to Mutual Gains?* Instruments for Sustainable Private Sector Forestry Series, International institute for Environment and Development, London

Medina, G. (2004) 'Big trees, small favors: Loggers and communities in Amazonia', *Bois et Forêts des Tropiques*, no 282

Medina, G., Pokorny, B. and Campbell, B. (2007) 'Institutional restrictions faced by Amazonian communities for managing their forests', *ETFRN News*, pp47–48

Medina, G., Pokorny, B. and Weigelt, J. (2009) 'The power of discourse: Hard lessons for traditional forest communities in the Amazon', *Forest Policy and Economics*, vol 11, pp392–397

Miller, R. P. and Nair, P. K. R. (2006) 'Indigenous agroforestry systems in Amazonia: From prehistory to today', *Agroforestry Systems*, vol 66, pp151–164

Miller, R. P., Penn, J. W. and Leeuwen, J. V. (2006) 'Amazonian homegarden: Their ethnohistory and potential contribution to agroforestry development', in B. M. Kumar and P. K. R. Nair (eds) *Tropical Homegarden: A Time-tested Example of Sustainable Agroforestry*, Springer, Dordrecht

Milz, J. (1998) *Guía para el Establecimiento de Sistemas Agroforestales*, manual, La Paz, Bolivia

MINAG and INRENA (2006) 'Plan Nacional de Reforestación (2005–2024)', in *Resolución Suprema*, MINAG, Lima

Ministerio del Ambiente (2006) 'Plan nacional de forestación y reforestación (Acuerdo Ministerial)', MINAG, Lima

Mohan, S., Alavalapati, J. R. R. and Nair, P. K. R. (2006) 'Financial analysis of homegarden: A case study from Kerala state, India', in B. M. Kumar and P. K. R.

Nair (eds) *Tropical Homegarden. A Time-tested Example of Sustainable Agroforestry*, Springer, Dordrecht

Montagnini, F. and Nair, P. K. R. (2004) 'Carbon sequestration: An underexploited environmental benefit of agroforestry systems', *Agroforestry Systems*, vol 61–62, pp281–295

Mussak, M. and Laarman, J. (1989) 'Farmers' production of timber trees in the cacao-coffee region of coastal Ecuador', *Agroforestry Systems*, vol 9, pp155–170

Nair, K. S. S. (1993) 'State of the art agroforestry research and education', *Agroforestry Systems*, vol 23, pp95–119

Nair, K. S. S. (2001) *Pest Outbreaks in Tropical Forest Plantations: Is There a Greater Risk for Exotic Trees Species?*, CIFOR, Bogor, Indonesia

Nair, P. K. R. (1987) 'Agroforestry systems in major ecological zones of the tropics and subtropics', ICRAF Working Paper No. 47, in ICRAF/WMO International workshop on the application of meteorology to agroforestry systems planning and management, Nairobi

Navarro, R. M., Henríquez, N. C and Cornejo, J. A. (2005) *The Economic and Social Context of Monoculture Tree Plantations in Chile: The Case of the Commune of Lumaco, Araucania Region*, World Rainforest Movement, Montevideo

Nawir, A., Santoso, L. and Mudhofar, I. (2002) 'Towards mutually beneficial partnerships in outgrower schemes: Lessons learned from Indonesia. PLT Programme, Plantation Forestry on Degraded or Low-Potential Sites', in CIFOR (ed) *Equitable Partnerships between Corporate and Smallholder Partners*, CIFOR, Bogor, Indonesia

Nawir, A. A., Murniati, L. R. and Rumboko, L. (eds) (2007) *Forest Rehabilitation in Indonesia: Where To After Three Decades* CIFOR, Bogor, Indonesia

Nepstad, D. C., Uhl, C. and Serreo, A. S. (1991) 'Recuperation of a degraded Amazonian landscape: Forest recovery and agriculture restoration', *Ambio*, vol 20, pp248–255

Neto, P. J. S., Melo, A. C. G. and Costa, F. C. M. (2004) 'Avaliação do sistema agroflorestal cacaueiro (*Theobroma cacao* L.) e Mogno (*Swietenia macrophylla* King) em Medicilândia, PA', in *Proceedings of the V Concresso Brasileiro de sistemas agroflorestais. 'SAFs: Desenvolvimento com proteção ambiental'* SBSAF (Sociedade Brasileira de Sistemas Agroflorestais), Curitiba

Nilsson, B. (2003) *Establecimiento de un programa de plantaciones forestales en Bolivia: Posibilidades y requerimientos*, Asdi, Scandiaconsult Natura, Cámara Forestal de Bolivia, Santa Cruz, Bolivia

Niskanen, A. (2000) 'Forest plantations in the South: Environmental–economic evaluation', in M. Palo and H. Vanhanen (eds) *World Forests from Deforestation to Transition?*, Kluwer Academic Publisher, Dordrecht

Padoch, C., Chota Inuma, J., de Jong, W. and Unruh, J. (1985) 'Amazonian agroforestry: A market-oriented system in Peru', *Agroforestry Systems*, vol 3, pp47–58

Pandey, D. and Ball, J. (1998) 'The role of industrial plantations in future global fibre supplies', *Unasylva*, vol 193

Pattanayak, S. K. and Mercer, D. E. (2002) 'Indexing soil conservation: Farmer perceptions of agroforestry benefits', *Journal of Sustainable Forestry*, vol 15

Peck, R. B. (1982) 'Actividades de investigación en bosques e importancia de los sistemas de multiestratos en la cuenca Amazónica (neotrópicos húmedos)', in S. B. Hecht (ed.) *Amazonia: Investigación sobre agricultura y uso de tierras*, Centro Internacional de Agricultura Tropical (CIAT), Cali, Colombia

Peck, R. B. (1990) 'Promoting agroforestry practices among small producers: The case of the Coca Agroforestry project in Amazonian Ecuador', in A. B. Anderson (ed.) *Alternatives to Deforestation Steps Toward Sustainable Use of the Amazonia Rain Forest*, Columbia University Press, New York

Pichon, F. J. (1997) 'Colonist land-allocation decisions, land use, and deforestation in the Ecuadorian Amazon frontier', *Economic Development and Cultural Change*, vol 45, pp707–744

Pinedo-Vasquez, M., Zarin, D. J., Coffey, K., Padoch, C. and Rabelo, F. (2001) 'Post-boom logging in Amazonia', *Human Ecology*, vol 29, pp219–239

Pokorny, B. and Johnson, J. (2008) 'Community forestry in the Amazon: The unsolved challenge of forests and the poor', policy brief, ODI Natural Resource Perspectives no 112, ODI, London

Pokorny, B. and Phillip, M. (2008) 'Certification of NTFP. Concluding comments', *Forests, Trees and Livelihoods*, vol 8(1), pp91–95

Puri, S. and Nair, P. K. R. (2004) 'Agroforestry research for development in India: 25 years of experiences of a national program', *Agroforestry Systems*, vol 61, pp437–452

Raik, D. B. and Decker, D. J. (2007) 'A multisector framework for assessing community-based forest management: Lessons from Madagascar', *Ecology and Society*, vol 12, p14

Ramirez, A., Sere, C. and Uquillas, J. (1992) 'An economic analysis of improved agroforestry practices in the Amazon lowlands of Ecuador', *Agroforestry Systems*, vol 17, pp65–86

Rice, R. and Greenberg, R. (2000) 'Cacao cultivation and the conservation of biological diversity', *Ambio*, vol 29, pp167–173

Rogers, E. M. (2003) *Diffusion of Innovations*, Free Press, New York

RRI/ITTO (Rights and Resources Initiative/International Tropical Timber Organization) (2009) *Tropical Forest Tenure Assessment: Trends, Challenges and Opportunities*, RRI and ITTO, Yokohama

Satterthwaite, D. and Sauter, G. (2008) *Understanding and Supporting the Role of Local Organisations in Sustainable Development*, IIED Gatekeeper Series, no 137, IIED, London

Sayer, J., Chokkalingam, U. and Poulsen, J. (2004) 'The restoration of forest biodiversity and ecological values', *Forest Ecology and Management*, vol 201

Scherr, J. S., White, A. and Kaimowitz, D. (2004) *A New Agenda for Forest Conservation and Poverty Reduction: Making Markets Work for Low Income Producers*, Forest Trends, Washington, DC

Schirmer, J. and Tonts, M. (2003) 'Plantations and sustainable rural communities', *Australian Forestry*, vol 66, pp67–74

Schmerbeck, J. (2003) *Patterns of Forest Use and its Influence on Degraded Dry Forests: A Case Study in Tamil Nadu, South India*, Shaker Verlag, Aachen

Sears, R., Padoch, C. and Pinedo-Vasquez, M. (2007) 'Amazonia forestry transformed: Integrating knowledge for smallholder timber management in Eastern Brazil', *Human Ecology*, vol 35, pp697–707

Shanley, P. and Medina, G. (2005) *Frutíferas e Plantas ùteis na Vida Amazônica*, CIFOR and Imazon, Belém

Shanley, P., Luz, L. and Swingland, I. R. (2002) 'The faint promise of a distant market: A survey of Belém's trade in non-timber forest products', *Biodiversity and Conservation*, vol 11, pp615–636

Simmons, S., Walker, R. T. and Wood, C. H. (2002) 'Tree planting by small producers in the tropics: A comparative study of Brazil and Panama', *Agroforestry Systems*, vol 56, pp89–105

Smith, N. J. H., Falesi, I. C., Alvim, P. D. T. and Serrão (1996) 'Agroforestry trajectories among smallholders in the Brazilian Amazon: Innovation and resiliency in pioneer and older settled areas', *Ecological Economics*, vol 18, pp15–27

Summers, P. M., Browder, J. O. and Pedlowski, M. A. (2004) 'Tropical forest management and silvicultural practices by small farmers in the Brazilian Amazonia: Recent farm-level evidence from Rondônia', *Forest Policy and Economics*, vol 192, pp161–177

Sunderlin, W. D., Angelsen, A., Belcher, B., Burgers, P., Nasi, R., Santoso, L. and Wunder, S. (2005) 'Livelihoods, forests, and conservation in developing countries: An overview', *World Development*, vol 33, pp1383–1402

Tamale, E., Jones, N. and Pswarayi-Riddihough, A. (1995) *Technologies Related to Participatory Forestry in Tropical Forestry in Tropical and Subtropical Countries*, World Bank Technical Paper No. 289, World Bank, Washington

Teixeira, F. A. M. (2005) *Economia do cacau na Amazônia*, UNAMA, Belém

Thacher, T., Lee, D. R. and Schelhas, J. W. (1996) 'Farmer participation in reforestation incentive programs in Costa Rica', *Agroforestry Systems*, vol 35, pp269–289

UN (2002) *Report of the World Summit on Sustainable Development Johannesburg, South Africa, 26 August–4 September 2002*, United Nations, New York

UN (2008) *The Millennium Development Goals Report*, United Nations Department of Economic and Social Affairs, New York

UNCED (1992) 'Agenda 21: Deforestation', in *United Nations Conference on Environment and Development, June 3–14, 1992*, United Nations, New York

UNDCP (1997) *Manejo, conservación y utilización de los recursos forestales en el Trópico de Cochabamba y en las zonas de transición de los Yungas de La Paz – Fase II*, UNDCP, New York

Varmola, M. I. and Carle, J. B. (2002) 'The importance of hardwood plantations in the tropics and sub-tropics', *International Forestry Review*, vol 4, pp110–121

Vidal, N. G. and Donini, G. (2004) *Promising business models for Community–Company Collaboration in Brazil and Mexico*, 'Proceedings of the Tenth Biennial Conference of the International Association for the Study of Common Property (IASCP). The Commons in an Age of Global Transition: Challenges, Risks and Opportunities', IASCP, Oaxaca

Vivan, L. L., Monte, N. L. and Gavazzi, R. A. (2002) 'Implantación de tecnologías de manejo agroforestal en tierras indígenas del Acre', in *Experiêcias PDA*, Comisión Pro-Indio del Acre

WBCSD (2001) *Aracruz Celulose: The Forestry Partners Program. Case Studies*, World Business Council for Sustainable Development, Geneva

Weyerhaeuser, H., Wen, S. and Kahrl, F. (2006) *Emerging Forest Associations in Yunnan, China: Implications for Livelihoods and Sustainability*, International Institute for Environment and Development, Edinburgh, UK

Wiersum, K. F. (1982) 'Tree gardening and taungya on Java: Examples of agroforestry techniques in the humid tropics', *Agroforestry Systems*, vol 1, pp53–70

Wightman, K. E., Cornelius, J. P. and Ugarte-Guerra, L. J. (2006) 'Plantemos madera! Manuel sobre el establecimiento, manejo y aprovechamiento de plantaciones maderables para productores de la Amazonía peruana', in ICRAF (ed.) *Manual Técnico 04*, ICRAF, Lima

WRM (1999) *Pulpwood Plantations: A Growing Problem*, World Rainforest Movement, Montevideo

Xu, D. (2003) *Scenarios for a Commercial Eucalypt Plantation Industry in Southern China: Eucalypts in Asia, Zhanjiang*, ACIAR, Guangdong, People's Republic of China

7

Policies to enhance the provision of ecosystem goods and services from plantation forests

Peter J. Kanowski

Introduction

The need to develop policies that enhance the provision of ecosystem goods and services from plantation forests has emerged in parallel with the rise of plantation forests as both a land use and source of industrial wood supply. The history of plantation forestry has been characterized by policies that focused on expanding the plantation estate for wood production rather than for a wider set of values and outcomes. Many of the plantation forests established under these policies replaced native or managed ecosystems with little regard for the wider consequences, and in doing so have frequently diminished, rather than enhanced, the output of ecosystem goods and services at both landscape and stand scales. There are also, however, contrary examples, which are consistent with an ecosystem approach to sustainability and the principles of sustainable forest management.

The mix of the public and private, and priced and unpriced, ecosystem goods and services associated with plantation forests, and the spatial and temporal variability associated with many of them, mean that developing policies to enhance their provision is particularly challenging. The emergence of 'new generation' approaches to environmental and sustainability policy is particularly relevant to the provision of ecosystem goods and services from plantation forests. These approaches suggest that a governance regime that recognizes the roles of both state and non-state actors, and which draws in situation-specific terms on a mix of regulatory, voluntary, economic and community-based instruments and mechanisms, is most likely to enhance the provision of ecosystem goods and services from plantation forests. These regimes must recognize the broader policy contexts which shape land-use, investment, production and conservation decisions, by both large- and small-scale forest growers. Market-based instruments and mechanisms are an important component of governance regimes to enhance the provision of ecosystem goods and services from plantation forests, but the role of the state remains fundamental, and voluntary and community-based initiatives can also be important. Effective systems to govern forest practices and monitor the delivery of ecosystem goods and services are also necessary for these regimes to succeed. Governance and management regimes must be adaptive, recognizing that policy development and implementation are essentially experimental, and thus incorporate timely monitoring, evaluation and revision.

A suite of principles and design criteria relevant to governance regimes to enhance the provision of ecosystem goods and services from plantation forests emerge from both theory and practice. They suggest a logical series of steps for planning, developing and managing plantation forests if the provision of ecosystem goods and services is to be enhanced; these steps are embedded in policy learning and adaptive management frameworks. The steps involve: the development of relevant knowledge, of both ecosystems goods and services and their social values; reaching societal agreement about the contributions which plantation forests should make to the overall provision of ecosystem goods and services, and establishing planning and management processes to realize those goals; designing and implementing governance regimes for ecosystem goods and services; and establishing effective *forest practices systems* and systems for monitoring the delivery of ecosystem goods and services.

As other chapters of this book, and other reviews (e.g. Cossalter and Pye-Smith, 2003; Kanowski, 2003; Kanowski and Murray, 2008) have discussed, plantation forests can impact both favourably and adversely on the provision of ecosystem goods and services. The ecosystem goods and services associated with plantation forests are – like those of forests more generally – a mix of the public and private, and the priced and unpriced; they are also delivered, to varying degrees, differentially and interdependently between and across landscapes and tenures, and at a variety of spatial and temporal scales (Maginnis and Jackson, 2003; Mercer, 2005). These characteristics of goods and services from forests are well recognized as being particularly challenging for policy development and implementation (Cubbage et al, 1993; Mayers and Bass, 1999; Mercer, 2005), and well suited to the suite of 'new generation' approaches to environmental and sustainability policy (e.g. Dovers, 2005; Gunningham, 2007) that have emerged in response to the limitations of previous policy regimes.

This chapter reviews policy learning from experience with plantation forests, and with ecosystem goods and services, to establish the relevant policy contexts. It then reviews relevant features of 'new generation' approaches to environmental and sustainability policy, as the basis for establishing a framework in which to situate policies to enhance the provision of ecosystem goods and services from plantation forests. Finally, it draws from recent literature to discuss how such policies might be developed and implemented within this framework.

The historical context – plantation forestry policies in the 20th century

The dramatic and continuing expansion of plantation forests, which has taken place largely over the past century and accelerated over the past 40 years, has occurred in response to a range of factors: growing global demand for forest products; market and technological forces favouring plantation-grown wood; a diminishing supply from native forests as a consequence both of overexploitation and of public policy decisions to reserve forests from harvesting; the declining attractiveness or competitiveness of other land uses; and the adoption of policies to promote plantation forests (Cossalter and Pye-Smith, 2003; Kanowski, 2005; Weber, 2005). Policies promoting plantation expansion have been widely adopted at national and sub-national levels since the 1920s (Mather, 1993; Kanowski, 2001; Garforth and Mayers, 2005), and sometimes at regional scales (e.g. Europe – Weber, 2005; Southern Africa – SADC, 2002 in Christy et al, 2007).

Around 50 per cent of 20th-century plantation forests were established for industrial wood production; the balance being established for fuelwood, land restoration or amenity (FAO, 2001). Since the 1980s, an increasing proportion of plantations has been established and intensively managed for industrial wood production (Kanowski, 2005). Thus, whilst the delivery of environmental benefits motivated the establishment of some plantation forests, industrial wood has increasingly become the dominant rationale for their establishment (Mather, 1993). This chapter is therefore concerned principally with policies directed at enhancing the contributions of these commercially oriented plantation forests. Other literature (e.g. CIFOR, 2006; Chazdon, 2008) focuses specifically on plantation afforestation for non-commercial objectives.

Largely as a consequence of the rationale for their establishment, most 20th-century plantation forests can be described as 'simple' (*sensu* Kanowski, 2001), in terms of intent, composition, structure and management regime. Many of these simple plantation forests have been outstandingly successful in terms of their economic objectives: wood yields per unit area and time are many orders of magnitude greater than those from natural forests, plantations are providing an increasing proportion of the world's wood supply, and – in many plantation regions – plantation-based economic development is demonstrable, significant and sustained (Kanowski, 2003, 2005). However, the impact of these simple plantation forests on ecosystem goods and services – such as biodiversity, landscape values or water quality – has often been adverse rather than favourable, particularly but not only where they have replaced natural or semi-natural ecosystems (e.g. Chapters 4 and 5; Kanowski, 2001; Cossalter and Pye-Smith, 2003). The social impacts of these plantation forests have also been mixed and context-specific: while plantations have generated economic development and employment, some have also been established in ways which accorded little attention to the rights and interests of indigenous and traditional owners and the poor, or which were not sensitive to their wider social context. In these cases, their social impacts have been, or have the potential to be, adverse (e.g. Chapter 6; Carrere and Lohman, 1996; Cossalter and Pye-Smith, 2003; Schirmer, 2005; Weber, 2005; Colchester, 2006; World Rainforest Movement, 2007).

Sustainable forest management – with its recognition of environmental and social, as well as economic, goals – is now the accepted basis of forest policy and practice globally (e.g. World Forestry Congress, 2003; Sayer and Maginnis, 2005; FAO, 2007a, 2007b). Thus, 21st-century plantation forests – and the policies that shape them – are expected to deliver environmental and social benefits, or at a minimum not cause environmental or social harm, as well as delivering the economic benefits on which they have historically focused. There are also self-interested, as well as more noble, reasons for plantation forest growers to adopt sustainable forest management principles; there is ample evidence from many countries (e.g. Australian and Irish case studies reviewed by Schirmer, 2005; Indonesia – Global Carbon Project et al, 2006; various South American countries – World Rainforest Movement, 2005) that meeting environmental and social expectations is necessary for plantation-based industries to maintain the *social licence* to operate, which they require to succeed over the decadal timeframes that typify large-scale plantation growing and processing investments.

Policy lessons from a century of plantation forestry

The experiences of the last century of plantation forestry are instructive in helping to shape its future. Reviews by, amongst others, Mather (1993), Cossalter and Pye-Smith (2003), Garforth and Mayers (2005), Kanowski (2005) and Weber (2005) suggest a number of general conclusions relevant to the topic of this chapter. These are summarized below.

Plantation policy is important, but only in context

The expansion of the plantation forest resource over the past century was not serendipitous; rather, it was the consequence of policies deliberately adopted by national and sub-national governments to promote plantation forestry, principally for wood production and the associated economic and social benefits. Such policies continue to be strong elements of some countries' forest policies (e.g. Australia – Plantations2020, 2007; Indonesia – Ministry of Forestry, 2005, in World Bank, 2006; South Africa – Department of Water Affairs and Forestry (DWAF), 2006).

However, as the reviews above also illustrate, plantation policy objectives have been facilitated, or hindered, by a variety of other forces, which may be more significant than 'plantation policies' themselves – including levels of supply and demand and the extent of trade, the relative attractions of competing land uses, the attitudes of landowners to tree growing, the degree to which governments were willing to use financial or other mechanisms to promote plantation forestry, and the commitment to policy goals and measures over extended periods. At a more operational level, factors such as levels of investment, management skills, and the availability of appropriate technologies have also been significant in shaping the economic, environmental and social outcomes of plantations.

Together, these experiences demonstrate the importance to plantation development of thorough and comprehensive policy analysis, of policy coordination and integration, of consistent policy directions sustained beyond the short-term, and of flexible and adaptive strategies to give effect to policy directions.

Governments play a number of important roles

Governments have played a central role in the expansion of plantation forests over the past century. They have done so either by direct public investment and involvement – for example in Australia, China, India, New Zealand or the UK – or by facilitating private sector investment, variously through financial policy instruments and other mechanisms such as facilitating access to land – for example in Brazil, Chile, Indonesia, Thailand, or South Africa (Mather, 1993; Garforth and Mayers, 2005; Kanowski, 2005). In most countries, governments have played both direct and facilitating roles; and in almost all of those once dominated by public forestry enterprises, the role of the private sector has progressively increased (Mather, 1993; Garforth and Mayers, 2005; Kanowski, 2005).

The facilitating role of government focused historically on addressing disincentives to private sector investment and participation in plantation forestry growing and product processing; for example, governments have typically sought to minimize the disincentive imposed by the cost structure of plantation growing, by providing various forms of financial incentive; they have often invested in the infrastructure and technological development necessary to catalyse private sector investment in the plantation sector; they have funded research to understand landowner attitudes to tree growing, and how greater adoption of plantation forestry might be encouraged. Subsequently, governments have also shaped and set the rules for markets for ecosystem goods and services. In summary, government's role in setting the framework for private sector investment and behaviour is as fundamental to plantation forestry as it is in other arenas of economic activity and environmental policy.

Policies need to recognize the importance of small-scale growers to the future of plantation forestry

Given the capital and scale requirements necessary to develop plantation resources adequate for competitive forest industries, most 20th-century plantation programmes were initiated by government agencies or large corporate entities (Mather, 1993; Garforth and Mayers, 2005). Subsequently, constraints to the roles of these larger actors, and the demonstrated and potential roles which small-scale private sector growers can play in association with others, has focused attention on the role of small-scale growers (e.g. Chapter 6; Mayers and Vermeulen, 2002). Whilst it can be challenging to develop tree-growing partnerships which are mutually satisfactory, and which contribute adequately to small-scale growers' livelihoods (e.g. Mayers, 2006), a range of cases (e.g. Desmond and Race, 2002; Mayers and Vermeulen, 2002) demonstrate it is possible – given supportive institutional and partnership arrangements, and a favourable operating context.

Engaging with small-scale growers is also becoming more important for the future of the plantation forestry sector for a range of other reasons – for example, as land ownership fragments (e.g. in the US; Brown, 2006), where the majority of land available for plantation establishment is that of small-scale landowners, as is now commonly the case (Mayers and Vermulen, 2002; Kanowski, 2003); or where policy objectives focus on small-scale landowners (e.g. Australia – Australian Greenhouse Office, 2003; Indonesia – Ministry of Forestry, 2005, in World Bank, 2006; South Africa – DWAF, 2006). Small-scale growers are more likely than industrial-scale growers to integrate commercial tree growing with agricultural production, in systems that are often described as 'agroforestry' or 'farm forestry', highlighting the need for a policy environment which supports such integration (e.g. Byron, 2001). Some forms of 'community forestry' may similarly be oriented towards commercial tree growing, and similarly require an enabling policy environment (e.g. Vermeulen et al, 2008).

Policies need to recognize that plantations can generate substantial environmental and social costs, as well as benefits

Over the past century, plantation policy-makers and growers have given considerable attention to the financial costs and benefits of plantations (e.g. Grayson, 1993; Kelly et al, 2005), principally to inform strategies to promote and sustain plantation forestry. However, it is clear that plantation forestry can also generate substantial environmental and social costs, as well as benefits (Kanowski, 2005). Like the financial costs, these are context-specific; they have been the subject of reviews at scales ranging from the local (e.g. Routley and Routley, 1973, for south-eastern Australia), to the national (e.g. Stewart, 1987, for the UK) and the transnational (e.g. Carrere and Lohman, 1996; World Rainforest Movement, 2007).

These costs comprise both direct and opportunity costs; they were seldom the focus of plantation policies, but were – explicitly or implicitly – judged by decision-makers to be costs worth bearing in pursuit of plantation-based development. Many critics of plantation forestry development (e.g. those cited above) argue that such judgements have been misplaced, and that approaches with different starting points – which respect customary rights, the principle of prior and informed consent, and the environmental values of land proposed for plantation development – are necessary to address the interests of parties other than plantation proponents (e.g. Colchester, 2006). The principles of sustainable forest management demand that these potentially adverse impacts be addressed more adequately in the future than they were in much of the last century of plantation forestry expansion.

Effective forest practice systems are important, across all tenures

Formal forest practice codes were not generally developed until the 1980s. Prior to this, there were few formal constraints on forest practices associated with establishing, managing and harvesting plantations, and the poor land management practices that often resulted led to many adverse impacts on ecosystem goods and services, and in turn to community discontent with plantations (e.g. Routley and Routley, 1973; Stewart, 1987; Mather, 1993).

Many governments responded by introducing codes of forest practice, or the equivalent (e.g. Dykstra and Heinrich, 1996), but their implementation has often been inadequate (Forsyth, 1998); as Gunningham (2007) notes, such 'implementation deficits' often characterize natural resource management more generally. It is also common for such codes to apply differentially to public and private tenures, and perhaps to different forms of forestry and to large- and small-scale operations (McDermott et al, 2007). Where forest practices systems have been implemented effectively, they have been successful in mitigating the adverse environmental impacts of forest operations (e.g. Enters et al, 2002; Wilkinson, 2003) – demonstrating their fundamental importance to the realization of sustainable forest management across all tenures and scales of operation.

With enabling policies and progressive management, plantations can deliver enhanced environmental and social benefits

Many progressive plantation growers have already sought to address the potentially adverse environmental and social impacts of plantation forestry, informed by research on both its environmental (e.g. Chapters 3, 4 and 5) and social (e.g. Bureau of Rural Sciences, 2005) outcomes and how they might be enhanced. A striking feature of the findings of much of this research is that substantial gains in environmental and social outcomes can be often made at relatively little direct or opportunity cost to economic returns, given adequate planning (e.g. Carnus et al, 2003; Lindenmayer et al, 2003; Schirmer, 2005).

Examples of progressive responses include, but are not limited to, the independent environmental and social assessments commissioned by some companies to inform their plans (e.g. Stora Enso in China; UNDP China, 2005), the transformation of public forestry agencies' approach to plantations (e.g. UK Forestry Commission; Grundy, 2005), and the development and implementation of multi-function plantations as public–private partnerships (e.g. Western Australia; Future Farm Industries CRC, 2007). Many of these responses are consistent with the emerging emphasis on the role of plantations in forest landscape restoration (e.g. Rietbergen-McCracken et al, 2006; IUCN, 2007), and with more 'complex' forms of plantation forestry, which I have argued previously (e.g. Kanowski and Savill, 1992; Kanowski 2001) are necessary to realize the potential of plantation forests to deliver environmental and social benefits. It is clear that approaches to conceiving, planning and managing plantation forests which are more sophisticated and balanced than those adopted for many simple plantations (see Chapter 5 for examples) can deliver greatly enhanced environmental and social benefits at tolerable economic cost.

It is helpful to reconceptualize plantation forestry for the 21st century

Historically, plantations have been defined as those comprising planted trees of particular species composition, age structure and spacing; that terminology is now evolving to the more inclusive term of 'planted forests' (FAO, 2007a), recognizing the diversity of forms, composition, scale and goals of these forests. Consistent with this evolution, this chapter considers 'plantation forests' in the broad sense – as 'trees established on a sufficient scale, and with income generation ranking sufficiently high as one of the purposes, to have reasonable potential as a commercial crop' (Kanowski, 2003) – rather than in terms of spatial configuration, species composition or management. This definition emphasizes the commercial goal of plantation forestry, but does not imply a preference for any particular model of tree growing; rather, it allows a diversity of forms and scales, ranging from traditional 'simple' plantation forestry to a variety of more complex systems, including those commonly referred to as 'agroforestry' (Kanowski, 2001).

The potential impacts of plantation forests on ecosystem goods and services

The ecosystem goods and services associated with forests have been reviewed by, amongst others, Landell-Mills and Porras (2002), Scherr et al, (2004), the Millennium Ecosystem Assessment (2005), Wunder (2005) and earlier chapters in this book. A comprehensive list (e.g. Daisy, 1997, in Scherr et al, 2004; Millennium Ecosystem Assessment, 2005) would include the role of forests in carbon, hydrological and nutrient cycles; in local climate regimes; in sustaining soil and catchment values; in sustaining biodiversity and related functions such as pollination services and pest control; and in the beauty of landscapes or other cultural values (Chapter 2).

The impacts of any particular plantation forest on ecosystem goods and services may vary, spatially and temporally, reflecting its landscape context, its design and composition, and its management; relevant issues have been reviewed in other chapters of this book. Examples of positive impacts include the use of plantation forests to restore connectivity in fragmented landscapes, to sequester atmospheric carbon, or to mitigate soil erosion. Examples of adverse impacts include loss of biodiversity or aesthetic value associated with landscape-scale conversion from natural ecosystems to plantation forests, reduced catchment water yields (e.g. Waterloo et al, 1999) or poor forest management practices impacting adversely on soil structure and water quality (e.g. Scott et al, 2005).

Policies addressing the impacts of plantation forests on ecosystem goods and services must therefore consider two sets of issues; the first are those associated with avoiding or mitigating the adverse impacts, and the second with delivering the potential benefits, of plantation forests for ecosystem goods and services. In the case of potentially adverse impacts, the issues separate into two categories; those associated with the choices about land use, and those associated with the management of plantation forests.

Of the adverse impacts, the dominant issue is the extent to which conversion from natural or semi-natural ecosystems to plantation forest is acceptable. This issue remains strongly contested. Positions range from opposition to any further conversion (e.g. for Indonesian peatlands – Global Carbon Project et al, 2006; Tasmania – The Wilderness Society, 2004; WWF Australia, 2004), to reluctant acceptance that conversion is likely to continue to proceed, especially in countries with high proportions of relatively natural ecosystems (FAO, 2007a), and a focus on protecting forests or other natural ecosystems with high conservation or cultural value (e.g. WWF, 2002). Whilst there may also be important impacts on ecosystem goods and services associated with the replacement of other land uses by plantation forests (e.g. changes in water yields when pastureland is converted to forest – Vertessy et al, 2003), the impacts associated with the conversion of ecosystems of high conservation or cultural values are likely to be the most profound and contentious.

Policies addressing land-use change to plantations already exist in many countries, and voluntarily through some forest certification systems. Some governments (e.g. New Zealand) and certification systems (e.g. the Forest Stewardship Council) preclude any conversion of natural to plantation forests; others allow conversion in particular land-use classes (e.g. Indonesia), or to specified thresholds (e.g. Brazil). In some cases (e.g. South Africa), conversion from grassland to plantation may also be precluded by concerns about impacts on water yield. These policy measures have been reviewed, for the most important plantation-growing countries, by McDermott et al (2010).

The second set of adverse impacts are those associated with the design, composition and management of plantation forests once land-use decisions have been made. These issues are determined largely by the goals of the plantation owner, and the opportunities and constraints under which they are operating, as well as by the particular environmental context. These impacts, which are associated more with operational than with land-use choice issues, are addressed – in a variety of ways, and to varying degrees – by both forest practices and forest certification systems (McDermott et al, 2010).

Turning to policies which might enhance ecosystem goods and services, rather than avoid or mitigate adverse impacts on them, the issue of the baseline from which 'enhance' is defined is a critical policy issue (e.g. Wunder, 2005). A number of perspectives are relevant. From an ecologist's perspective, the optimum levels of ecosystem goods and services might be those associated with functional natural systems, and 'enhance' might be defined as moving towards those levels. In contrast, a natural resource manager might seek to increase the yields of particular ecosystem goods and services without prejudicing others; for example, by thinning forests enough to enhance water yield, but not so much as to adversely impact on biodiversity or landscape values (e.g. Chapter 5). The difficulties encountered with defining 'additionality' in the context of forest-related activities under the Kyoto Protocol of the UNFCCC (e.g. IPCC, 2000) illustrate both the complications in agreeing a definition of 'enhance', and how disabling this can be for policies seeking to enhance the provision of ecosystem goods and services from forests. Governments are beginning to adopt pragmatic approaches to this issue; for example, the California Climate Action Registry's Forest Protocols (CCAR, 2007) defines the baseline for an ecosystem good (in this case, carbon) as the stocks and flows resulting from an approved management plan with a 100-year planning horizon, i.e., effectively 'those beyond the legal minimum that could be delivered'. Although they can be economically inefficient, by rewarding some actors unnecessarily, such pragmatic definitions are likely to be necessary to establish markets for ecosystem goods and services.

As many – both proponents and critics of plantation forests – have pointed out, plantations are not the equivalent of other forms of forests (e.g. Carrere, 2004; American Tree Farm System, 2007). There are roles in the delivery of economic, environmental and social benefits that plantation forests can play, and others which are better delivered by other forms of forests and forestry. The challenge to policies seeking to enhance the provision of ecosystem goods and services from plantations is to identify and capitalize on the comparative advantage of plantations in particular landscape and social contexts, whilst ensuring that the delivery of benefits does not prejudice sustainable forest management principles and their local interpretation at national and sub-national levels. This will often require that plantations be situated, in policy as well as practical terms, in a landscape context (Maginnis and Jackson, 2003; Sayer and Maginnis, 2005; Rietbergen-McCracken et al, 2006).

The next section discusses how contemporary thinking about the 'new environmental governance' might inform the development of such policies.

Contemporary thinking about environmental governance regimes

As the magnitude and urgency of environmental challenges, at scales from the local to global, have become more apparent, both policy-makers and civil society have focused on the formulation and implementation of policy responses. In recognition of the interdependent economic, environmental and social dimensions of environmental policy, many of these responses have been framed in terms of the broader issue of sustainability (Dovers, 2005), and in terms of 'environmental governance' – a terminology which encompasses environmental policy but explicitly recognizes the role of non-state actors in determining environmental and sustainability outcomes (Gunningham, 2007; CIFOR, 2009).

This broadening of the frame of reference for environmental policy is appropriate for a number of reasons. A principal one is that approaches to environmental policy have followed broader policy trends over the past three decades; in the most general and oversimplified terms, this trend has been away from policy regimes which emphasize the role of the state and of regulation, often characterized as 'command and control', to more heterogeneous and broadly based regimes which are collectively described as the 'new environmental governance' (Gunningham, 2007). This evolution was prompted both by the rise of neo-conservative governments, with their 'small government' agenda, since the 1980s, and by the evident limitations of traditional environmental regulation (Gunningham and Grabosky, 1998). Thus, the 'new environmental governance' emphasized the role of actors other than government, of processes such as voluntary and negotiated agreements, of approaches involving 'regulatory flexibility', and of market-based instruments (e.g. Gunningham and Sinclair, 2002; Cashore et al, 2004; Gunningham, 2007).

Although 'the new governance ... is defined more by what it is not, than by what it is' (De Burca and Scott, 2006, in Gunningham, 2007), Gunningham (2007) suggests, nevertheless, that it can:

> be treated as involving a cluster of characteristics: participatory dialogue and deliberation, devolved decision-making, flexibility rather than uniformity, inclusiveness, transparency, institution-alized consensus-building practices, and a shift from hierarchy to *heterarchy*. Not all these characteristics need to be present for a particular experiment to be regarded as involving new environmental governance, but the more characteristics that are present, and the stronger the form in which they are present, the greater is the claim to be regarded as falling within this category.

As a new generation of environmental governance has evolved, it has come to recognize the role, as well as the limitations, of the state; in particular, it acknowledges the fundamental role of the state in establishing and enforcing the rules within which other actors operate (Gunningham, 2007). Conversely, the lack of capacity or will of many governments to enforce laws and implement regulations provides one of the principal catalysts for non-state-based approaches to environmental governance. The case of largely inconclusive intergovernmental attempts to address forest loss and degradation since the 1992 Earth Summit (reviewed by Humphreys, 2006), and the associated emergence of forest certification (reviewed by Cashore et al, 2004) provide a good example of this phenomenon in the forests context.

Similarly, other market-based instruments, such as payments for environmental services, have emerged as one of the primary tools of new environmental governance regimes (e.g. Landell-Mills and Porras, 2002; Wunder, 2005; Sanchirico and Siikamäki, 2007). As Gunningham (2007) notes, the new environmental governance 'provides greater scope for non-state actors to assume administrative, regulatory, managerial and mediating functions previously undertaken by the state'. This scope is well illustrated in the forest context by aspects of both forest certification (e.g. Cashore et al, 2004) and the environmental services markets already established for some forests (e.g. Wunder, 2005).

Gunningham (2007) observes that new environmental governance arrangements are generally still young and dynamic, and are already evolving as their strengths, limitations and costs – in both absolute terms and relative to state-based approaches – become more apparent. Thus, various forms of 'regulatory pluralism' (Gunningham, 2007) are emerging in environmental governance – recognizing that the best combination of state and non-state governance mechanisms will depend on the particular context, and that these may vary over time, as well as between jurisdictions and particular environmental policy issues. In the best cases, these regimes are emerging in a consciously adaptive way, as the result of deliberate policy learning (Dovers, 2005). In many cases, the policy challenges are substantial, and the policy issues 'wicked', in the sense that there are few 'win–win' outcomes (Dovers, 2005; Gunningham, 2007). It is in this complex, dynamic and challenging milieu that policies, and governance arrangements more broadly, to enhance the provision of ecosystem goods and services from plantation forests are situated.

A governance framework for ecosystem goods and services from plantation forests

The 'new environmental governance' summarized above provides the basis for articulating a governance framework relevant to the provision of ecosystem goods and services from plantation forests. A number of such frameworks have already been developed for particular forest-related issues. For example, Binning and Young (2000) presented a systems model for native vegetation conservation, and subsequently developed it to focus on environmental services from farm forestry (Binning et al, 2002; van Bueren et al, 2002). Their model is structured around three core categories, represented as regulatory, economic and people. Similarly, Harrison et al (2003) presented a comparable, albeit differently worded, framework for 'policy instruments for environmental protection or to promote remedial action', which they used to situate environmental services payments for small-scale forestry. Wunder (2005, 2006) presents a framework for situating payments for environmental services in the context of other policy initiatives for forest conservation and sustainable forest management, and Scherr and White (2002) provide an overview of the policy instruments that might be used to promote environmental services from forests, and discuss – as have others (e.g. Kousky, 2005) – the factors that might be used to choose amongst them.

The framework developed here, illustrated in Figure 7.1, draws from these precursors – and Dovers' (2005) representation of the stages of policy development, which includes those of problem-framing, policy-framing and policy implementation – and adapts them in the light of Gunningham's (2007) review. The core components of the framework are the mechanisms and instruments – regulatory, voluntary, economic and community-based – which form the basis for constructing, in pluralistic and context-specific ways, the governance arrangements appropriate and feasible for particular environmental issues. Gunningham (2007) presents a number of examples from outside the forestry sector – in pollution and natural resource management – which illustrate how such arrangements have been developed and have evolved. As with these examples, governance regimes relevant to the delivery of ecosystem goods and services from plantation forests need to accommodate the roles of both the public and the private sectors, and large- and small-scale actors, if they are to be effective.

The framework also explicitly recognizes both the contexts created by other policies and institutional arrangements relevant to plantations and ecosystem goods and services, which can range from enabling to disabling, and the systems by which operational forest practices and the delivery of environmental goods and services are implemented and monitored, as fundamentally important to the outcomes of plantation forestry, including its provision of ecosystem goods and services.

The framework presented in Figure 7.1 also seeks to make explicit both the interactions and the complementarities between its core components; van Bueren et al (2002) note that 'policy mixes which harness the synergies between ... [components] ... are more likely to be effective (both in terms of cost and environmental outcomes) than those which use only single instruments'. Similarly, Mercer (2005, citing work by Gottfried et al, 1996) points out that economic theory demonstrates that 'market forces [alone] in decentralized, unregulated economies are inadequate for optimizing ecological services at a landscape level'.

Components of the governance framework

This section discusses each of the components of the governance framework presented in Figure 7.1, and relates them to the provision of ecosystem goods and services from plantation forests.

Policy and institutional contexts

Typically, a suite of policies and institutional arrangements at levels ranging from the international to the local – some specifically concerned with forests, others not – will define the context and policy 'space' for policies and

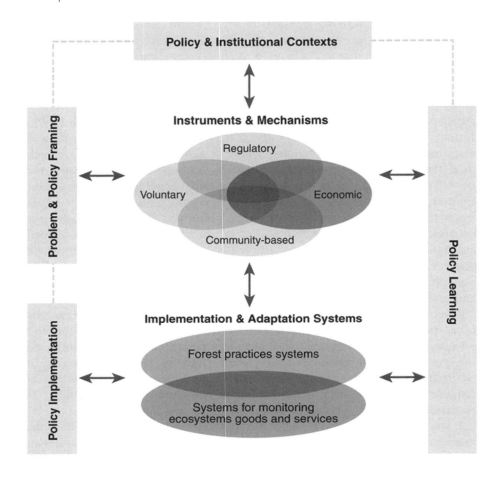

Figure 7.1 *A governance framework for ecosystem goods and services from plantation forests*

institutional arrangements focused on particular sectors or issues. These, of course, also reflect prevailing political ideologies – for example, an emphasis on market forces rather than on regulation. They may also be shaped by goals proposed by non-state actors that have acquired standing, such as IUCN's targets for the extent of protected area representation (World Commission on Protected Areas (WCPA), 2005).

Policy consistency, both vertically and horizontally, and the realization of synergies between different policies and institutional arrangements – internationally, nationally and at the sub-national level – are fundamental to realizing intended policy outcomes (Dovers, 2005). This is often challenging – as the example of the relationship between the ecosystem approach, adopted by the Convention on Biological Diversity, and the approaches to sustainable forest management promoted by the United Nations Forum on Forests and other multilateral forests processes, discussed by Sayer and Maginnis (2005), demonstrates. The purpose of the discussion here is not to review international arrangements for forests (see, e.g. Humphreys, 2006; Christy et al, 2007), but rather to comment on their principal implications for policy design at the national and sub-national levels.

International policy regimes

As Christy et al (2007, Chapter 2) point out, 'though there is little international law that directly governs forestry ... a number of international instruments do affect the forestry options of a country that is party to them'. On the basis of their analysis, the Convention on Biological Diversity, the Framework Convention on Climate Change and its Kyoto Protocol, the UN Convention to Combat Desertification, and trade agreements such as the International Tropical Timber Agreement and those adopted under the auspices of the World Trade Organization, emerge as the multilateral agreements most generally and strongly relevant to plantation forests; others, such as the Ramsar Convention on the Protection of Wetlands, the Convention on International Trade in Endangered Species of Wild Species of Flora and Fauna, and the (Bonn) Convention on the Conservation of Wild Species of Migratory Animals, will be relevant in specific circumstances. They note the potential of treaties addressing the rights of Indigenous Peoples, principally the 1989 ILO (International Labour Organization) Convention No 169 Concerning Indigenous and Tribal People in Independent Countries, and the 2007 UN Declaration on the Rights and Responsibilities of Indigenous Peoples, to shape forest-related policy and practice. Regional and bilateral treaties – such as the 1993 Central American Forests Convention, the 2002 Southern African Development Community Protocol on Forestry, or various European Union conventions, declarations and regulations – are also important in defining contexts for their member states.

In the absence of a forests convention, non-legally binding agreements are particularly relevant to forestry activities. The UN Forum on Forests' 2007 agreement of 'a non-legally binding agreement on all types of forests', and its Programme of Work and Plan for Action (UNFF, 2007), represent the principal multilateral agreement; however, the various processes to define and report on sustainable forest management are also important (FAO, 2007b), and arguably more immediately relevant to plantation forests and the goods and services they generate.

In summary, the two dominant themes that emerge from the international policy context are the commitments of countries, established since the 1992 Earth Summit, to the interrelated goals of forest conservation and sustainable forest management, and – to varying degrees – to mechanisms to enable them. The principal implication for plantation forests is that their establishment and management should support, not diminish, these goals. More recently, as climate and energy issues rise up the political agenda, there has been renewed focus on the role of plantations in carbon sequestration (e.g. Chapter 3; Kerr et al, 2004) and as sources of renewable energy (e.g. European Union, 2007). Whilst these agendas help to create a new policy space for plantation forests and ecosystem goods and services, they also present significant challenges in designing national and sub-national policies which reconcile and advance these three themes in mutually supportive ways.

National and sub-national policy regimes

Given the diversity and complexity of individual national and sub-national contexts and arrangements, generalizations are useful only in terms of higher-level principles. Nevertheless, a consistent shortlist of policy characteristics emerges from both general reviews relevant to environment and forest policy (e.g. Cubbage et al, 1993; Mayers and Bass, 1999; Gunningham and Sinclair, 2002; Dovers, 2005; Christy et al, 2007) and those which focus more specifically on ecosystem services, plantations or sustainable forest management (e.g. Scherr and White, 2002; Garforth and Mayers, 2005; Sayer and Maginnis, 2005; Stanturf and Madsen, 2005). National and sub-national policy regimes more likely to enable positive environmental and sustainability outcomes will be characterized by:

- consistency of policy goals with those agreed internationally;
- consistency and synergy of plantation-focused policies with other national and sub-national policies;
- clearly defined tenure and use rights for both forests and ecosystem goods and services;
- good governance, including appropriate public participation processes;
- adequate institutional capacity, in both the public and private sectors;
- functional institutional arrangements, which are purposive and persistent, but also sufficiently adaptive and flexible;
- an explicit policy learning approach.

National and sub-national policy regimes also need to deliver the core elements of sustainable forest management relevant to plantation forests and ecosystem goods and services. These comprise:

- land-use planning processes and mechanisms which ensure:
  - the conservation of ecosystem goods and services at the appropriate landscape scale, which will vary: for example, bio- (i.e. eco-)regional planning is recognized as the best basis for conservation planning (Kanowski et al, 1999; The Nature Conservancy, 2006), whereas catchments are the relevant scale for planning related to water yield (Vertessy et al, 2003);
  - conservation at the stand scale where necessary, such as for forests of high conservation or cultural value;

- systems which assess and monitor, at appropriate scales and levels of precision, the delivery of environmental goods and services from landscape components, including plantation forests (e.g. Mercer, 2005; Alpízar et al, 2007);
- forest practices systems which are effective at the landscape and stand scales, and across tenures and management regimes (e.g. Wilkinson, 2003);
- policy mechanisms and instruments which are effective, efficient and adaptive, and relevant to both large- and small-scale plantation growers (e.g. Mercer, 2005);
- an explicit adaptive management approach (e.g. Lindenmayer and Franklin, 2005).

These features are consistent with those suggested by WWF (2002) as key elements of sustainability for plantation forests; WWF's elements also include social criteria, viz. respect for the rights of communities and indigenous peoples, and positive social impacts.

The particular mix of policy mechanisms and instruments which best deliver these outcomes will vary between jurisdictions and with the target ecosystem goods and services, as discussed in the following section.

Policy instruments and mechanisms

The instruments and mechanisms from which governance regimes for particular purposes can draw are classified here, following Gunningham (2007), as regulatory, voluntary, economic and community-based. Apart from ideologues who believe in either the unfettered role of markets or the incontestable role of the state, there is a high degree of consensus (see, e.g. Scherr and White, 2002; van Bueren et al, 2002; Mercer, 2005; Gunningham, 2007) that environmental governance regimes should draw as appropriate on each of the range of instruments and mechanisms, in forms that reflect the particular political, institutional, social and economic circumstances, to deliver their goals in efficient and effective ways. As Mercer (2005) observes in the context of forests and ecosystem services, 'deciding on the best combination of policies ... requires determining which ecosystem services are amenable to market solutions, which require government intervention, and which require a combination of government and market approaches'; one could extend his list to include voluntary and community-based approaches.

The following sections briefly review each of the categories of instrument and mechanism, and how they form part of environmental governance regimes relevant to ecosystem goods and services from plantation forests.

Regulatory instruments and mechanisms

Two suites of regulatory instruments and mechanisms are relevant to this topic. The first comprises those focused on land and forest management, which 'use laws and policies to either dictate specific land management actions or otherwise limit or control how landowners (public and private) manage their lands' (Mercer, 2005). As discussed above, the limitations of these regulatory approaches – principally their actual and political costs and feasibility, and their relative effectiveness and efficiency – have been largely responsible for the emergence and adoption of other approaches to environmental governance. Nevertheless, regulatory instruments and mechanisms continue to play a fundamental role in shaping, to varying degrees depending on the jurisdiction, central policy issues – such as where plantations can and cannot be established, the conservation obligations of and constraints on landowners, or the minimum standards required of forest operations; these in turn impact on the ecosystem goods and services outcomes from plantation forests.

The second suite of regulatory instruments and mechanisms comprise those with which governments shape markets for ecosystem services, the space for community-based initiatives, and – perhaps to a lesser extent – that for voluntary actions. These policy settings, and the willingness of governments to intervene should they fail to deliver intended outcomes, play a significant role in determining the success of these other elements of environmental governance regimes (Gunningham, 2007). For example, the baselines set by governments are crucial to the success or otherwise of ecosystem services markets (e.g. Wunder, 2005), as are the authority and legitimacy which governments allow for community-based forest management (e.g. review and results in Pagdee et al, 2006).

Voluntary instruments and mechanisms

Voluntary approaches to environmental governance are appealing for a variety of reasons – for example, as an alternative to politically difficult regulation, because they appeal to landowners' commitment to stewardship, because of their greater flexibility and lower costs (Mercer, 2005), or because they may fit with progressive business strategies (Gunningham and Sinclair, 2002). However, there are also real limits to voluntarism – it will not necessarily address priorities for environmental outcomes (Mercer, 2005), and it is unlikely to succeed in curtailing activities with adverse environmental outcomes unless private and public interests coincide (Gunningham, 2007).

Nevertheless, voluntary approaches to delivering ecosystem goods and services from forests are important and significant. Examples relevant to plantation forests include the conservation easements common in the US, including those associated with the sale of forestlands (e.g. The Nature Conservancy, 2007), the growth of voluntary carbon markets (Capoor and Ambrosi, 2007), or the commitment of individual forest owners or businesses to identifying and protecting high conservation forests (e.g. Mondi, 2007). In some cases, forest certification systems may require plantation design and management practices that deliver ecosystem goods and services: examples include the protection and enhancement of remnant natural forest in Brazil, or of native grassland and water yield in South Africa (May, 2006; Kanowski and Murray, 2008).

As Mercer (2005) notes, governments frequently encourage voluntary actions directed at forest conservation and ecosystem restoration by providing economic incentives of various forms, as discussed below.

Economic instruments and mechanisms

Two sets of economic instruments and mechanisms are widely used to enhance the delivery of ecosystem goods and services from forests. The first comprises direct financial incentives from public funds to private forest owners; these include direct payments, or those through cost-sharing arrangements or subsidies, or tax relief. These mechanisms are widely used in, and often fundamental to, natural resource policy regimes seeking to enhance the delivery of ecosystem goods and services (e.g. Scherr et al, 2004; Mercer, 2005; Weber, 2005). For example, most payments for water catchment-related services are transfer payments from downstream users to upstream land managers (Scherr et al, 2004); cost-sharing with landowners is the basis of many schemes for biodiversity conservation on private lands, such as the state of Victoria's BushTender (Department of Sustainability and Environment, Victoria, 2007).

The second set of instruments and mechanisms function through the operation of markets for ecosystem goods and services, which now exist for a number of ecosystem goods and services, in a variety of forms and scales (see reviews by Landell-Mills and Porras, 2002; Scherr et al, 2004; Wunder, 2005; Bracer et al, 2007). The value of these markets, while small – in comparison to the value of the principal forest product valued by markets, viz. timber, and to their potential – is nevertheless significant; for example, Scherr at al (2004) estimated the annual value of direct and indirect payments for ecosystem services of tropical forests to approximate that of total annual investment in tropical forest conservation, viz. about US$2.5 billion.

Scherr et al (2004) classify these market mechanisms as, in addition to the direct payments discussed above, 'open trading under a regulatory cap or floor; self-organized private deals; and ecolabelling of forest or farm products, an indirect form of payment for ecosystem services'. Wunder (2005) suggests a complementary classification based on three distinctions: between area-based schemes, such as conservation easements, and product-based schemes, such as ecolabelling; between public and private schemes, noting that the former are generally larger in scope and more ambitious in goals; and between use-restricting schemes, such as those focused on conservation, and those focused on asset-building, such as reforestation of deforested sites.

The most developed and most international of the trading markets are those for carbon sequestration (Scherr et al, 2004), although the proportion of forest-based carbon traded remains small, at around 1 per cent of total carbon market volume, due to regulatory complexities and limited access to the European market (Capoor and Ambrosi, 2007). Trading markets have also been established for a few other environmental services in some countries, e.g. wetland mitigation banking in the US (Scherr et al, 2004). In the case of biodiversity, Landell-Mills and Porras (2002) reported that, although various payment mechanisms for biodiversity conservation have been developed, markets remain largely nascent and experimental, reflecting both the intangibility and diffuseness of biological diversity. Similar constraints apply to markets for landscape beauty (Landell-Mills and Porras, 2002). In some cases, the 'bundling' of ecosystem goods and services offers a means to overcome some of these limitations, and to reduce transaction costs (Landell-Mills and Porras, 2002; Wunder, 2005).

There is little information about private arrangements for ecosystem goods and services, but these are thought to be relatively limited in extent and impact (Scherr et al, 2004). In contrast, the ecolabelling of forest and farm products has developed rapidly (e.g. Cashore et al, 2004; Scherr et al, 2004), and in some cases has made significant impacts on forest management for ecosystem goods and services – although its benefits to producers have usually been in terms of market access rather than in premium returns (Scherr et al, 2004; Wunder, 2005).

Community-based instruments and mechanisms

Community-based instruments and mechanisms are particularly relevant where natural resources are under community management (e.g. Wunder, 2005; Pagdee et al, 2006), but may also be important where community-based movements emerge for conservation or land rehabilitation (e.g. Australia's Landcare – Landcare Australia, 2007; Kenya's Green Belt Movement – Green Belt Movement, 2007). Use of community-based approaches may be a conscious political choice to empower communities (e.g. Robinson, 2004, in Gunningham, 2007), or may reflect the established authority, effectiveness and legitimacy of community-based management (e.g. Borrini-Feyerabend et al, 2008).

There are examples of communities self-organizing, or with external facilitation, and securing payments for watershed protection and for carbon sequestration, typically through reforestation of various forms (Scherr et al, 2004; Wunder, 2005). The structure of most ecosystem goods and services markets currently favours – in Wunder's (2005) terminology – asset-building rather than use-restricting activities (Scherr et al, 2004), and thus facilitates the provision of ecosystem goods and services from plantations. Scherr et al (2004) note that rewarding the stewardship role which many communities play in conserving forests will require the development of market structures more amenable to supporting use-restricting activities.

Systems to govern forest practices and the delivery of environmental goods and services

Mayers and Bass (1999) and Bass (2001) characterized the situation during the 1990s, in which environmental and forest policies were multiplying but institutional and operational capacity diminishing, as one of 'policy inflation, capacity collapse'. This dysfunctional paradox highlights the need for effective systems for operational implementation and monitoring, both of forest management (Wilkinson, 1999) and for the delivery of ecosystem goods and services (Scherr and White, 2002; Mercer, 2005; Wunder, 2005). Such systems are also fundamental to the adaptive management approaches inherent in sustainable forest management (e.g. Lindenmayer and Franklin, 2005), and underpin policy learning approaches to environmental governance (e.g. Dovers, 2005).

Forest practices systems were reviewed briefly above. Wilkinson (2003) notes the importance of establishing forest practices systems within an appropriate regulatory framework; he identifies the components of this framework as laws and policies, strategic and operational planning processes, implementation and enforcement arrangements, monitoring systems, and review and improvement processes. As with environmental governance regimes more generally, forest practices systems can vary in emphasis along a spectrum from command and control to self-regulation (Wilkinson, 2002); Wilkinson (2003) discusses the advantages in a co-regulatory approach, consistent with a new environmental governance model. Similarly, systems to measure and monitor the delivery of ecosystem goods and services from landscape components are essential in order to design, implement and assess the effectiveness and efficiency of the policy initiatives intended to enhance them (Binning et al, 2002; Alpízar et al, 2007). The state of development of these systems varies (Wunder, 2005); for example, they are now reasonably well developed for carbon sequestration, which has the advantage of being relatively generalizable, but more embryonic for water catchment services or biodiversity conservation, which are more spatially and perhaps temporally variable. Extending these systems to assess the social values of outcomes is especially challenging, but is necessary if investment choices are to be well informed and unbiased (Mercer, 2005).

Designing governance regimes to enhance the provision of ecosystem goods and services from plantation forests

As noted in the preceding discussion, governance regimes to enhance the provision of ecosystem goods and services from plantation forests need to be specific to particular policy, institutional and implementation contexts. They will also need to be adaptive, recognizing that policy development and implementation are essentially experimental (Dovers, 2005), and thus incorporate adequate monitoring, evaluation and review processes. In this context, a number of recent reviews (Landell-Mills and Porras, 2002; Scherr and White, 2002; Scherr et al, 2004; Gunningham, 2007) have suggested principles or criteria to inform the development of these governance regimes and the choice of policy instruments and mechanisms. The discussion below draws from these.

The role of the state

A fundamental issue in designing governance regimes is that of the role of the state, as this defines – deliberately or by default – the space for, and role of, other elements of new-generation approaches. Gunningham (2007) suggests that a number of state roles are clearly evident from experience to date with new-generation approaches to environmental governance:

- definitional guidance – describing and defining the nature of the governance arrangements; including their focus, intended outcomes, geographical and participatory scope, legal basis, funding arrangements, and relationship to other institutional structures. In many cases, this will also extend to a policy coordination role once arrangements are implemented;
- participatory incentives – either positive inducements or negative sanctions for targeted actors to participate in the particular governance arrangements;
- enforcement capability – to ensure that other participants fulfil their obligations, particularly where – as is commonly the case – there are costs to some actors of compliance.

A second set of roles which governments play are those noted by Scherr et al (2004) – in establishing the enabling conditions and setting the rules for markets for ecosystem goods and services, and in engaging in those markets as both a buyer of many ecosystem services and a catalyst for private-sector participation. These activities include establishing baselines and caps, and regulations that govern trading schemes; establishing or clarifying property rights; and facilitating the development of the institutions necessary for markets to function efficiently and with acceptable transaction costs.

Choosing instruments and mechanisms

Scherr and White (2002) identify five key factors – implementation contexts, biophysical features of ecosystem services, management complexity, economic costs, and equity – as fundamental to the choice of instruments and mechanisms to enhance the provision of ecosystem goods and services. Each of these is discussed below.

Implementation contexts

The general implementation contexts are established by the international, national and sub-national policy regimes and institutional arrangements discussed above. Within these, Scherr and White (2002) identify the features of political, institutional and economic conditions directly relevant to the choice of instruments and mechanisms to promote ecosystem services.

The political conditions comprise:

- the extent of environmental consciousness in society;
- the perceived legitimacy of different interests and forms of action; and
- the relative power of those different interests.

The institutional conditions comprise:

- the clarity and security of rights to ecosystem services and to land;
- the capacity of local institutions; and
- the degree of trust that exists between key stakeholders.

The economic conditions comprise:

- the perceived economic value of the ecosystem service, and the associated willingness of users to pay for the service;
- the capacity of producers of the service to respond to payment with improved management for that service; and
- the sensitivity of the instrument to opportunity costs.

Interpreting these in general terms, it is clear that higher levels of environmental consciousness are likely to empower both the state and other actors seeking environmental outcomes, but outcomes will still be shaped by the relative power and interests of different groups and societal mores about the legitimacy of different approaches – such as, about the degree of regulation versus voluntary action. For example, forest practices systems for private forests in the south-eastern US have a much greater degree of voluntarism than those in the Pacific coast states of the US (McDermott et al, 2007).

The institutional conditions described by Scherr and White (2002) each represent enabling conditions; the greater each is, the more likely that effective governance arrangements to enhance the provision of ecosystem goods and services can be realized, and that those arrangements can be built around non-regulatory instruments and mechanisms. Similarly, where an ecosystem service is highly valued, users have a high willingness to pay, payments will strongly impact on management, and the instrument is sensitive to opportunity costs, some form of market-based mechanism is likely to be attractive and effective in enhancing ecosystem goods and services. Delivery of high-quality drinking water from forested catchments might be such an example (Porras et al, 2008).

Biophysical features of ecosystem goods and services

The purpose of characterizing the biophysical features of ecosystem goods and services is to establish the nature of relationships between them and land-use and management regimes (Scherr and White, 2002; Wunder, 2005). As discussed above, and as many reviews of forest ecosystem services (e.g. Binning et al, 2002; Landell-Mills and Porras, 2002; Scherr et al, 2004; Mercer, 2005; Wunder, 2005) note, there are many complexities and various uncertainties associated with the measurement of ecosystem services and their attribution to particular parts of the landscape and to particular management regimes. In some cases, such as carbon sequestration, the services are relatively easily measurable, approaches relatively generalizable, and the impacts of management regimes reasonably clear; for others, particularly those with strong and dynamic spatial and temporal dimensions – such as catchment services and biodiversity – assessment and attribution may be very difficult at levels of resolution relevant to small-scale landowners in complex landscapes.

Our relatively poor knowledge of many of these relationships, and the high costs of acquiring better knowledge, means that proxies and approximations are commonly necessary, and favour the bundling of ecosystem goods and services (Landell-Mills and Porras, 2002; Wunder, 2005). The lack of such knowledge for ecosystem goods and services judged to be important also defines the research agenda to inform policy development and implementation (Mercer, 2005; Bracer et al, 2007). As Mercer (2005) notes, there is also a parallel social sciences research agenda necessary to inform our understanding of the relationships between ecosystem goods and services, and social values.

Management complexity

It is evident from reviews such as those by Dovers (2005) or Gunningham (2007) that effective environmental governance arrangements are inherently complex, and that realizing their potential poses challenges to all actors, in both the public and private sectors. It is also the case that arrangements typically become more complex over time, as governance systems and markets adapt to new circumstances and seek to address deficiencies. Wilkinson (2002, 2003) noted this phenomenon for forest practices systems, and – similarly – ecosystem services markets have become increasingly complex and sophisticated as they have developed (Scherr et al, 2004; Capoor and Ambrosi, 2007). These realities emphasize the need both to develop arrangements which are locally appropriate and functional, and for investment in capacity-building to enable participation in and evolution of these arrangements (Scherr and White, 2002; Scherr et al, 2004; Bracer et al, 2007). Whilst this need is often most acute for small-scale producers in economically less-developed countries, it is not confined to them.

Costs and risks

The direct and transaction costs and their distribution, and the levels of risk, associated with different governance arrangements are important considerations in the design and implementation of environmental governance regimes. For example, Mercer (2005) presents a comparison of the relative costs – classified as transaction, monitoring, enforcement, opportunity and political – associated with a suite of regulatory, economic and voluntary instruments and mechanisms to promote environmental restoration, and considers their distribution among different interests. Suites of risks are associated with the structure and operation of markets, and are greatest for those who are least well-informed; Scherr et al (2004) identify the need for both public and private sector investments in institution-building and innovation, to reduce transaction costs, which may be very high – especially for small-scale participants – and the levels of financial risk inherent in market participation. Wunder (2005) suggests a number of strategies for mitigating these risks for small-scale, poor, participants in ecosystem services markets, and emphasizes the importance both of understanding their constraints and preferences, and of adapting mechanisms to address these.

Equity

Equity considerations for governance regimes to promote ecosystem goods and services relate principally to the regulatory treatment of different actors, and the extent to which they can access economic instruments and mechanisms. Historically, regulatory regimes such as forest practices systems have often treated different actors – e.g. public and private sectors, large- and small-scale landowners, and natural and plantation forests – differently (e.g. McDermott et al, 2007); given the interdependencies of ecosystem services across landscapes, more consistent treatment should be one of the goals of environmental governance regimes.

Small-scale producers face particular difficulties in engaging with and benefiting fully from commodity markets (e.g. Mayers and Vermeulen, 2002; Harrison et al, 2003), or from ecolabelling schemes (e.g. Nussbaum and Simula, 2005), and this is also the case for markets for ecosystem goods and services. This situation is exacerbated for the poor, particularly in economically less-developed nations (Scherr et al, 2004; Wunder 2005). At worst, the livelihoods of the natural resource-dependent poor may be threatened if their interests are displaced by ecosystem service-oriented land-use activities; conversely, ecosystem services markets also offer a potential means to generate economic and environmental benefits for the poor (Scherr and White, 2002; Wunder, 2005). As Wunder (2005) points out, as well as comparative disadvantages likely to characterize small-scale landowners – typically high transaction costs, limited resources, and often insecure rights – they may also have advantages – in particular, lower opportunity costs than other actors. However, as Scherr et al (2004) note, proactive strategies will be necessary to ensure that the poor and small-scale landowners can access and benefit from ecosystem service markets, as has been the case with other market-based mechanisms such as forest certification (Nussbaum and Simula, 2005).

Implementing governance regimes to enhance the provision of ecosystem goods and services from plantation forests

Given the diversity of national and sub-national contexts of the ways in which environmental governance regimes can be constituted, and of landscapes in which plantation forests are situated, the preceding discussion has necessarily been at a high level of generalization. This concluding section suggests a series of steps by which the general principles and elements of the governance framework presented in Figure 7.1, and discussed above, can be implemented. It assumes – consistent with the above definition under 'policy lessons from a century of plantation forestry' – that plantation forests are established for commercial purposes, but recognizes that – while wood production is likely to be an important commercial product – other goods and services may also be marketable, and are not necessarily subsidiary to wood or to other forest products in terms of the owner's objectives or their economic value.

Step 1. Develop and communicate the knowledge base

The purpose of this step is to characterize – at least to a level sufficient for policy and market development – the relationships between landscape components, including plantation forests, and ecosystem goods and services, and the values which societies accord particular ecosystem goods and services. As Mercer (2005) notes, this information represents the supply and demand sides, respectively, of the ecosystem goods and services equation. This knowledge base will also need to address the impacts of alternative management regimes.

As discussed, these relationships for forests and carbon sequestration are now relatively well known and generalizable; those for water are also relatively well known in general terms and more specifically for some catchments; but whilst the principles for biodiversity and landscape values are relatively apparent, their expression is typically very context-specific, and more complicated by the interactions between landscape components. For practical purposes, various levels of assumption and approximation are likely to be necessary.

This information – and its communication to policy-makers, forest managers and other stakeholders – should provide the basis for decisions about the roles that plantation forests will be expected to play in the delivery of ecosystem goods and services in particular landscape and social contexts.

Step 2. Agree what contributions plantation forests should make to the delivery of ecosystem goods and services, and establish plantation planning and management frameworks that deliver them

This step comprises two interdependent components. The purpose of the first is to reach sufficient societal consensus about the contributions plantation forests should make to the delivery of ecosystem goods and services in particular landscape and social contexts. This is an expression of social values, and is therefore inherently political; but it needs to be informed as much as possible by the scientific information generated by Step 1, and by the trade-offs likely to be identified in the planning processes described below.

The purpose of the second component is to establish plantation planning and management frameworks which give effect to these agreed goals, and to the principles of sustainable forest management. These processes should ensure that plantation establishment and management either avoid, or contain to agreed levels, the adverse impacts plantation forests can have on ecosystem goods and services; that the ecosystem goods and services expected, at landscape and stand scales, are delivered; and that these goals are realized consistent with the principles of sustainable forest management. In practice, this step requires conservation planning and management on a bioregional basis; planning and management on a catchment basis; landscape planning and management at appropriate, variable scales down to the level of forest stands; and the consideration of various land-use and management options. These planning and management processes will also need to respect forest and other landscape components of particular significance, such as those of high conservation or cultural value, and societal values more generally. They also require consideration of the form, composition and management of plantation forests that are appropriate and feasible in each particular context.

The outcomes of this step should be the identification of those parts of the landscape in which plantation forests could or should be established, their scale and design and their composition and management, so that they deliver the desired environmental goods and services as well as the intended wood and non-wood forest products.

Step 3. Design the governance regime for ecosystem goods and services, implement its components and monitor outcomes

The purpose of this step is to define and implement the most appropriate mix of regulatory, voluntary, economic and community-based instruments and mechanisms to deliver the outcomes sought from Step 2. The regime will be shaped by policy and institutional contexts, institutional capacities (by the extent to which markets exist or can be developed for ecosystem goods and services) and by social preferences and mores. Governments will play critical roles in shaping both regimes as a whole, and their components – for example, through the design and implementation of regulatory systems, through establishing the rules under which markets for ecosystem goods and services operate, and – where necessary – in catalysing and fostering markets, and voluntary and community-based initiatives. Both governments and business will need to pay particular attention to the impacts of regimes on small-scale landowners and forest growers, and the poor. There is now a wealth of literature, and an emerging body of practice, to inform and assist the development and evolution of appropriate governance regimes.

In the context of environment governance regimes, any particular instrument or mechanism is a means to an end, rather than an end in itself. However, in designing governance regimes, there may also be broader social and political objectives in the choice of instrument or mechanism: for example, fostering entrepreneurial skills, or strengthening community-based institutions.

The outcomes of this step should be a socially appropriate, functional and adaptive governance regime to enhance the provision of ecosystem goods and services from plantation forests of all scales.

Step 4. Establish effective forest practices systems, and systems for monitoring the delivery of ecosystem goods and services

The purpose of this step is to ensure that the operational management of plantations delivers the goals agreed above, and is consistent with the principles of sustainable forest management as articulated in forest practice codes or the equivalent. The essential features of forest practice systems, and the requirements for them to be implemented effectively, are now well established. These systems are the principal means for governing operational management, and thus ensuring that plantation forestry management activities comply with agreed expectations and standards, as well as providing essential information for adaptive management. Their effectiveness is likely to be enhanced by the participation of the forest grower in a credible forest certification scheme.

Similarly, the purpose of systems for monitoring the delivery of ecosystem goods and services is to ensure that agreed goals are realized. Their implementation has many commonalities with that of forest practice systems, and there are obvious advantages in integrating the two systems to the fullest extent possible.

The outcomes of this step will be plantation forest operations consistent with forest practice requirements, which deliver the ecosystem goods and services expected of them, and which are adaptive on the basis of information generated by constituent monitoring and review processes.

Conclusions

Because most plantation forests of the 20th century were narrowly conceived, and because the dominant interests in societies often accorded little formal value to the ecosystem goods and services on which plantation forests impacted, the impacts of many 20th-century plantation forests on ecosystem goods and services were adverse or sub-optimal. Changing societal values, more progressive thinking about the potential roles of plantation forests, and the emergence of new forms of environmental governance regimes intersect to create a policy environment which is much more enabling of the provision of ecosystem goods and services from plantation forests.

Consistent with the widely agreed principles of sustainable forest management, societies have the right to expect that plantation forests will deliver more benefits than costs – although the interpretation of how that balance is constituted will remain, ultimately, a value judgement. However, it is clear that enhancing the provision of ecosystem goods and services from plantation forests is an important element of realizing the benefits of this form of forestry, and gaining or maintaining social licence for plantation-based industries to operate. This is recognized by progressive plantation growers and by many forest certification schemes.

It is also apparent, from both operational experience and from research, that appropriately planned, designed and managed plantation forests can deliver a range of ecosystem goods and services, at both landscape and stand scales. The suite of impacts and benefits associated with plantation forests will typically differ from those of other landscape components, emphasizing the need to assess, agree and manage the impacts and contributions of plantation forests to ecosystem goods and services in a landscape context. The growing body of knowledge about these impacts and contributions provides the foundation for developing governance regimes which enhance, in ways consistent with the principles of sustainable forest management, the provision of ecosystem goods and services from plantation forests.

References

Alpízar, F., Blackman, A. and Pfaff, A. (2007) 'Payments for ecosystem services: Why precision and targeting matter', *Resources*, vol 165. pp20–22

American Tree Farm System (2007) 'What is tree farming?', www.treefarmsystem.org > About tree farming

Asia-Pacific Forestry Commission (1999) *Code of Practice for Forest Harvesting in Asia-Pacific*, FAO Regional Office for Asia and the Pacific, Bangkok

Australian and Tasmanian Governments (2005) *The Tasmanian Community Forestry Agreement: A Way Forwards for Tasmania's Forests*, The Commonwealth of Australia, Canberra and the State of Tasmania, Hobart

Australian Greenhouse Office (2001) 'The Contribution of mid to low rainfall forestry and agroforestry to greenhouse and natural resource management outcomes – overview and analysis of opportunities', www.greenhouse.gov.au/nrm/publications/index.html www.climatechange.gov.au

Bass, S. (2001) 'Policy inflation, capacity collapse', in R. J. Raison, A. G. Brown and D. W. Flinn (eds) *Criteria and Indicators for Sustainable Forest Management*, CABI, Wallingford

Binning, C. and Young, M. (2000) *Native Vegetation – Institutions, Policies and Incentives*, CSIRO Wildlife and Ecology, Canberra

Binning, C., Baker, B., Meharg,S., Cork, S. and Kearns, A. (2002) *Making Farm Forestry Pay – Markets for Ecosystem Services*, Rural Industries Research and Development Corporation

Bracer, C., Brana, J. and Swallow, B. (2007) 'Compensation and rewards for environmental services', www.ecosystemmarketplace.com, accessed 16 July 2007

Brown, M. (2006) 'What if there was a National Sustainable Forests Policy Act?', *Journal of Forestry*, vol 104, p341

Bureau of Rural Sciences (2005) *Socioeconomic Impacts of Plantation Forests*, Forest and Wood Products Research and Development Corporation, Melbourne

Byron, N. (2001) 'Keys to smallholder forestry in developing countries in the tropics', in S. R. Harrison and J. L. Herbohn (eds) *Sustainable Farm Forestry in the Tropics*, Edward Elgar, Cheltenham

Capoor, K. and Ambrosi, P. (2007) *State and Trends of the Carbon Market 2007*, The World Bank, Washington, DC

Carnus, J.-M., Parrotta, J., Brockerhoff, E. G., Arbez, M., Jactel, H., Kremer, A., Lamb, D., O'Hara, K. and Walters, B. (2003) 'Planted forests and biodiversity', www.maf.govt.nz/mafnet/unff-planted-forestry-meeting/index.htm

Carrere, R. (2004) 'Plantations are not forests', www.wrm.org.uy > Plantations Campaign > Briefings and documents, accessed 23 May 2010

Carrere, R. and Lohmann, L. (1996) 'Pulping the South', www.wrm.org.uy, accessed 23 May 2010

Cashore, B., Auld, G. and Newsom, D. (2004) *Governing through Markets: Forest Certification and the Emergence of Non-state Authority*, Yale University Press, New Haven

CCAR (California Climate Action Registry) (2007) 'Forest sector protocol', www.climateregistry.org, accessed 23 May 2010

Chazdon, R. L. (2008) 'Beyond deforestation: Restoring forests and ecosystem services on degraded lands', *Science*, vol 320, pp1458–1460

Christy, L., Di Leva, C. E., Lindesay, J. M. and Tella Tokoukam, P. (2007) *Forests Law and Sustainable Development: Addressing Contemporary Challenges through Legal Reform*, The World Bank, Washington, DC

CIFOR (2006) 'Review of forest rehabilitation initiatives: Lessons from the past', www.cifor.cgiar.org/rehab/, accessed 23 May 2010

CIFOR (2009) 'Forests and governance programme', www.cifor.cgiar.org/AboutCIFOR/How-we-work/Governance/Research.htm, accessed 23 May 2010

Colchester, M. (2006) 'Plantations for people?', *Arborvitae*, vol 31, p7

Cossalter, C. and Pye-Smith, C. (2003) *Fastwood Forestry – Myths and Realities*, CIFOR, Bogor, Indonesia

Cubbage, F. W., O'Laughlin, J. and Bullock, C. S. (1993) *Forest Resource Policy*, Wiley, New York

Department of Sustainability and Environment, Victoria (2007) 'BushTender', www.dpi.vic.gov.au/dse > Conservation and Environment > Biodiversity > Rural Landscapes > BushTender, accessed 23 May 2010

Department of Water Affairs and Forestry – RSA (2007) 'Water use licensing, registration and revenue collection', www.dwaf.gov.za/Projects/WARMS/, accessed 23 May 2010

Desmond, H. and Race, D. (2002) 'Global survey and analytical framework for forestry outgrower arrangements', in *Equitable Partnerships between Corporate and Smallholder Partners*, FAO and CIFOR

Dovers, S. (2005) *Environment and Sustainability Policy*, Federation Press

DWAF (Department of Water Affairs and Forestry), South Africa (2006) 'DWAF Multi-Year Strategic Plan 2007/8–2009/10; Programme 4 – Forestry', www.dwaf.gov.za/Documents/Other/Strategic%20Plan/MultiYear07.asp, accessed 23 May 2010

Enters, T., Durst, P. B., Applegate, G. B., Kho, P. C. S. and Man, G. (2002) *Applying Reduced Impact Logging to Advance Sustainable Forest Management*, FAO Regional Office for Asia and the Pacific, Bangkok

European Union (2007) 'Bioenergy: Objectives – technologies', www.ec.europa.eu/energy/res/sectors/bioenergy_en.htm

FAO (2001) 'Forest resources assessment 2000 – plantations', www.fao.org/forestry/site/10095/en

FAO (2007a) 'State of the world's forests 2007', www.fao.org/forestry/site/sofo/en/

FAO (2007b) 'What is sustainable forest management?', www.fao.org/forestry/site/24447/en/

FAO (2007c) 'Planted forests', www.fao.org/forestry/site/10049/en

FAO (2007d) 'Planted forests – definitions and concepts', www.fao.org/forestry/site/plantedforests/en/

Future Farm Industries CRC (2007) 'About us', www.futurefarmcrc.com.au/

Garforth, M. and Mayers, J. (eds) (2005) *Plantations, Privatisation, Poverty and Power*, IIED, London

Global Carbon Project, Global Environment Centre and CIFOR (2006) 'Riau Declaration on Peatlands and Climate Change', www.peat-portal.net/ev_en.php

Grayson, A. J. (1993) *Private Forestry Policy in Western Europe*, CAB International, Wallingford

Green Belt Movement (2007) 'Home page', www.greenbeltmovement.org/

Grundy, D. (2005) 'From plantation developer to steward of the nation's forests: The UK', in M. Garforth and J. Mayers (eds) *Plantations, Privatisation, Poverty and Power*, IIED, London

Gunningham, N. (2007) *The New Collaborative Environmental Governance*, International Meeting on Law and Society in the 21st Century, Humboldt University, Berlin

Gunningham, N. and Grabosky, P. (1998) *Smart Regulation: Designing Environmental Policy*, Clarendon Press, Oxford

Gunningham, N. and Sinclair, D. (2002) *Leaders and Laggards: Next Generation Environmental Regulation*, Greenleaf

Gunningham, N., Kagan, R. and Thornton, D. (2004) 'Social license and environmental protection: Why businesses go beyond compliance', *Law & Social Inquiry*, vol 29, pp307–341

Harrison, S., Killin, D. and Herbohn, J. (2003) 'Commoditisation of ecosystem services and other non-wood values of small plantations', in J. Suh, D. B. Smorfitt, S. R. Harrison and J. L. Herbohn (eds) *Marketing of Farm-grown Timber in Tropical North Queensland*, Rainforest CRC Research Report

Humphreys, D. (2006) *Logjam: Deforestation and the Crisis of Global Environmental Governance*, Earthscan, London

IPCC (2000) *Special Report on Land Use, Land-Use Change and Forestry*, Cambridge University Press, Cambridge

IUCN (2007) 'Forest landscape restoration', www.iucn.org/themes/fcp/experience_lessons/flr.htm

Kanowski, P. J. (2001) 'Plantation forestry at the millennium', in G. M. Woodwell (ed.) *Forests in a Full World*, Yale University Press, New Haven

Kanowski, P. J. (2003) 'Challenges to enhancing the contributions of planted forests to sustainable forest management', paper to UNFF Experts Meeting – Planted Forests, www.maf.govt.nz/mafnet/unff-planted-forestry-meeting/index.htm accessed 23 May 2010

Kanowski, P. (2005) 'Intensively managed planted forests', www.theforestsdialogue.org

Kanowski, P. and Murray, H. (2008) 'Intensively managed planted forests: Toward best practice', www.theforestsdialogue.org

Kanowski, P. J .and Savill, P. S. (1992) 'Forest plantations: Towards sustainable practice', in C. Sargent and S. Bass (eds) *Plantation Politics: Forest Plantations in Development*, Earthscan, London

Kelly, M, J., Tredinnick, J., Cutbush, G. and Martin, G. (2005) *Impediments to Investments in Long Rotation Timber Plantations*, Forest and Wood Products Research and Development Corporation, Melbourne, Australia

Kerr, S., Brunton, E. and Chapman, R. (2004) *Policy to Encourage Carbon Sequestration in Plantation Forests*, Motu Economic and Public Policy Research, Wellington

Kousky, C. (2005) 'Choosing from the policy toolbox', www.ecosystemmarketplace.com > Opinion, accessed 30 November 2005

Landcare Australia (2007) 'What is Landcare?', www.landcareonline.com/

Landell-Mills, N. and Porras, I. T. (2002) *Silver Bullet or Fools' Gold?*, IIED, London

Lindenmayer, D. B. and Franklin, J. F. (2005) *Towards Forest Sustainability*, CSIRO Publishing, Melbourne

Lindenmayer, D. B., Hobbs, R. J. and Salt, D. (2003) 'Plantation forests and biodiversity conservation', *Australian Forestry*, vol 66, pp62–66

McDermott, C. L., Cashore, B. and Kanowski, P. (2007) *A Global Comparison of Forest Practice Policies using Tasmania as a Constant Case*, Global Institute of Sustainable Forestry, Yale University, New Haven

McDermott, C. L., Cashore, B. and Kanowski, P. (2010) *Global Environmental Forest Policies*, Earthscan, London

Maginnis, S. and Jackson, W. (2003) 'The role of planted forests in forest landscape restoration', paper to UNFF Experts Meeting – Planted Forests, www.maf.govt.nz/mafnet/unff-planted-forestry-meeting/index.htm, accessed 23 May 2010

Mather, A. S. (1993) *Afforestation: Policies, Planning and Progress*, Belhaven Press

May, P. (2006) 'Forest certification in Brazil', in B. Cashore, F. Gale, E. Meidinger and D. Newsom (eds) *Confronting Sustainability: Forest Certification in Developing and Transitioning Countries*, Yale School of Forestry & Environmental Studies

Mayers, J. (2006) *Poverty Reduction through Commercial Forestry. What evidence? What prospects?*, The Forests Dialogue, Yale University Press, New Haven

Mayers, J. and Bass, S. (1999) *Policy that Works for Forests and People*, IIED, London

Mayers, J. and Vermeulen, S. (2002) *Company–Community Forestry Partnerships: From Raw Deals to Mutual Gains*, Instruments for Sustainable Private Sector Forestry series, IIED, London

Mercer, D. E. (2005) 'Policies for encouraging forest restoration', in J. A. Stanturf and P. Madsen (eds) *Restoration of Boreal and Temperate Forests*, CRC Press, Boca Raton

Millennium Ecosystem Assessment (2005) 'Overall synthesis', www.millenniumassessment.org > Synthesis reports

Mondi (2007) 'Sustainable forestry policy', www.mondigroup.com > Sustainability > Forestry

The Nature Conservancy (2006) *Conservation by Design*, TNC, Washington

The Nature Conservancy (2007) *Securing Innovative Conservation*, TNC, Washington

Nussbaum, R. and Simula, M. (2005) *The Forest Certification Handbook*, 2nd edn, Earthscan, London

Pagdee, A., Kim, Y. and Daugherty, P. J. (2006) 'What makes community forest management successful: A meta-study from community forests throughout the world', *Society and Natural Resources*, vol 19, pp33–52

Plantations2020 (2007) 'Plantations for Australia: The 2020 vision', www.plantations2020.com.au/

Porras, I., Grieg-Gran, M. and Neves, N. (2008) *All that Glitters: A Review of Payments for Watershed Services in Developing Countries*, IIED, London

Ramsar Convention Secretariat (2007) 'About Ramsar', www.ramsar.org

Rietbergen-McCracken, J., Maginnis, S. and Sarre, A. (2006) *The Forest Landscape Restoration Handbook*, Earthscan, London

Routley, R. and Routley, V. (1973) *The Fight for the Forests*, Research School of Social Science, Australian National University, Canberra

Sanchirico, J. N. and Siikamäki, J. (2007) 'Natural resource economics and policy in the 21st Century: Conservation of ecosystem services', *Resources*, vol 165, pp8–10

Scherr, S. and White, A. (2002) *Factors to Consider in Choosing Instruments to Promote Environmental Services*, presentation to conference on 'Payments for environmental services in China', Beijing, 22–23 April 2002

Scherr, S., White, A. and Khare, A. (2004) *For Services Rendered: The current status and future potential of markets for the ecosystem services provided by tropical forests*, ITTO, Yokohama

Schirmer, J. (2005) 'Achieving successful change in conflict over afforestation: A comparative analysis', PhD thesis, School of Resources Environment and Society, Australian National University, Canberra

Scott, D. F., Bruijnzeel, L. A. and Mackensen, J. (2005) 'The hydrological and soil impacts of forestation in the tropics', in M. Bonnell and L. A. Bruijnzeel (eds) *Forests, Water and People in the Humid Tropics*, Cambridge University Press, Cambridge

Stanturf, J. A. and Madsen, P. (2005) *Restoration of Boreal and Temperate Forests*, CRC Press, Boca Raton

Stephenson, K. (2009) 'Neither hierarchy nor network: An argument for heterarchy', *People and Strategy*, vol 32, pp4–13

Stewart, P. J. (1987) *Growing against the Grain*, Campaign to Protect Rural England, London

UNDP China (2005) *Environmental and Social Impact Assessment Analysis: Stora Enso Plantation Project in Guangxi, China*, UNDP China, Beijing

UNFCCC (2007) 'Kyoto Protocol', www.unfccc.int > Kyoto Protocol

UNFF (2007) 'About UNFF', www.un.org/esa/forests/about.html

van Bueren, M., Binning, C., Baker, B., Meharg, S., Kearns, A. and Cork, S. (2002) *Making Farm Forestry Pay – Selling the Environmental Services of Farm Forestry*, Rural Industries Research and Development Corporation, Canberra

Vermeulen, S., Nawir, A. A. and Mayers, J. (2008) 'Rural poverty reduction through business partnerships? Examples of experience from the forestry sector', *Environment, Development and Sustainability*, vol 10, pp1–18

Vertessy, R., Zhang, L. and Dawes, W. R. (2003) 'Plantations, river flows and river salinity', *Australian Forestry*, vol 66, pp55–61

Waterloo, M. J., Bruijnzeel, L. A., Vugts, H. F. and Rawaga, T. T. (1999) 'Evaporation from *Pinus caribaea* plantations on former grassland soils under maritime tropical conditions', *Water Resources Research*, vol 35, pp2133–2144

Weber, N. (2005) 'Afforestation in Europe: Lessons learnt, challenges remaining', in J. A. Stanturf and P. Madsen (eds) *Restoration of Boreal and Temperate Forests*, CRC Press, Boca Raton

The Wilderness Society (2004) *Protecting Forests, Growing Jobs*, Australia

Wilkinson, G. R. (1999) 'Codes of forest practice as regulatory tools for sustainable forest management', paper presented to 18th Biennial Conference of the Institute of Foresters of Australia, Hobart, 3–8 October 1999

Wilkinson, G. R. (2002) 'Building partnerships – Tasmania's approach to sustainable forest management', in T. Enters, P. Durst, G. B. Applegate, P. C. S. Kho and G. Man (eds) *Applying Reduced Impact Logging to Advance Sustainable Forest Management*, FAO RAP Publication, Rome

Wilkinson, G. (2003) 'Designing the regulatory framework for the implementation of codes of forest practice – the Tasmanian experience', paper for International Meeting on the Development and Implementation of National Codes of Practice for Forest Harvesting: Issues and Options, Kisarazu City, Japan, 17–20 November 2003

World Bank (2006) *Sustaining Economic Growth, Rural Livelihoods, and Environmental Benefits: Strategic Options for Forest Assistance in Indonesia*, The World Bank, Jakarata

World Commission on Protected Areas (2005) 'WCPA Strategic Plan 2005–2012', *World Commission on Protected Areas News*, vol 95, IUCN, Gland, Switzerland

World Forestry Congress (2003) 'Final statement: Forests, source of life', www.fao.org/forestry/site/32762/en/ > Twelfth World Forestry Congress

World Rainforest Movement (2005) 'Pulp mills: From monocultures to industrial pollution', www.wrm.org.uy > Plantations campaign > Pulp mills

World Rainforest Movement (2007) 'Plantations campaign', www.wrm.org.uy > Plantations campaign

Wunder, S. (2005) 'Payments for environmental services: Some nuts and bolts', CIFOR Occasional Paper No. 42, CIFOR, Bogor

Wunder, S. (2006) 'Are direct payments for environmental services spelling doom for sustainable forest management in the tropics?', *Ecology and Society*, vol 11, p23

WWF (2002) 'Forest plantations', position paper on plantations, May 2002, www.panda.org

WWF Australia (2004) *A Blueprint for the Forest Industry and Vegetation Management in Tasmania*, WWF Australia, Sydney

8
Ecosystem goods and services – the key for sustainable plantations

Jürgen Bauhus, Benno Pokorny, Peter J. van der Meer,
Peter J. Kanowski and Markku Kanninen

Improved ecosystem services from plantations of the future

Mankind has altered and transformed 40–50 per cent of the ice-free terrestrial surface of the Earth and appropriates an estimated 20 per cent of the global net primary production (Imhoff et al, 2004). The world's population is predicted to increase by 34 per cent from today to reach 9.1 billion people in 2050 (FAO, 2009). At the same time, this population will be more urbanized and have higher income levels than today. This larger and richer population will place unprecedented demands on the Earth's natural resources for the production of food and fibre (FAO, 2009). Currently humans appropriate 40–50 per cent of the available fresh water, and this is predicted to increase to 70 per cent by 2050 (Postel et al, 1996). In addition, it has been estimated that humans have doubled the nitrogen inputs from fertilizing agricultural systems and fossil fuel burning into terrestrial ecosystems (Vitousek et al, 1997), and the current atmospheric nitrogen deposition has enhanced the forest carbon sink by some 10–20 per cent (Schulze et al, 2009). These figures indicate that a large proportion of the increase in the production of food and fibre has come through intensification of agricultural and forest management systems, although extensification of agriculture and plantation forestry through land-use change, e.g. forest clearing, has also contributed to this. It is expected that 20 per cent of the required increase in food production in countries with developing economies between now and 2050 will be contributed through the extensification of agriculture. These projections demonstrate that there will be intense competition for land between different land-use options such as food production, production of timber, fibre and biofuels as well as nature conservation, and the development of urban areas and infrastructure. The growing demand for food, fibre, water, etc. is likely to lead to further deteriorations of ecosystem services as has been documented in the Millenium

Ecosystem Assessment (2005). The projections also stress that we need to be mindful about the possible consequences of replacing natural ecosystems that provide great benefits to human societies such as clean drinking water, soil protection, etc. Many of the ecosystem services they provide are irreplaceable, or the technology necessary to replace them is prohibitively expensive (Palmer et al, 2004). Thus the role of both natural and replacement, or man-made ecosystems in maintaining these services becomes increasingly important. In the past, tree plantations have had ambivalent roles with regards to ecosystem services. Their production function has served in a very efficient way to meet the growing demand for wood products. However, where tree plantations have replaced native ecosystems (forests or grasslands), many ecosystem services have deteriorated. The expansion of fast-growing industrial plantations for pulp, together with the rapid expansion of oil-palm plantations, has been a major driver of deforestation in the past, for example in Indonesia (Barr, 2002; Uryu et al, 2008). Where plantations were established on degraded or former agricultural land, many ecosystem services improved (see previous chapters and Cossalter and Pye-Smith, 2003). The future development of forest plantations, both in terms of extent as well as quality and management has to be viewed in this context.

Although there is great uncertainty about the future development of the spatial extent of the different land-use options (Lambin et al, 2001; Berndes et al, 2003; de Groot et al, in press), it is unlikely that tree plantations will occupy substantial shares of the land suitable for sustainable agriculture, unless there are political instruments such as a carbon tax, and incentives or subsidies, for example for bioenergy, that make the products and services from plantations more valuable than agricultural commodities (Johansson and Azar, 2007). Given the likely increasing competition for land and the growing need also for ecosystem services, plantations would be best established in locations where they contribute to improving ecosystem and landscape functioning and thus contribute to increasing the provisioning of ecosystem services. Due to improved industrial safeguards, certification schemes, REDD+ schemes, and needs for better corporate social responsibility, it is likely that much of the future expansion of plantations will take place in areas in need of restoration, i.e. degraded tropical lands (e.g. Otsamo, 2001). Also areas which are marginal for agriculture, or where synergies between agricultural production and tree plantations can be achieved may become more attractive for plantation expansion. This is particularly the case in those tropical countries, where the reduction of natural forest cover has produced large areas of forest and agricultural mosaics (de Jong, 2010).

Areas in which plantations can contribute positively to improving ecosystem services may comprise land degraded through unsustainable past land use (Evans and Turnbull, 2004; Metzger and Hüttermann, 2009), or marginal land, where agricultural intensification, e.g. through irrigation, is not possible. However, the term 'degraded land' is highly subjective and is not a terra nullius for afforestation projects. Livelihoods of many people, in particular in countries with developing economies, depend on the use of 'degraded' land, even if restoration through afforestation would yield higher returns in the future (Hunter et al, 1998). In the tropics and subtropics, much of the restoration of degraded land has been through afforestation with industrial monocultures involving a limited number of species from very few tree genera. Although these efforts may have been successful in terms of generating goods such as pulpwood, very few of these plantations provide the variety of goods and services to the local people that were once provided by the original forests or even the degraded systems that were replaced (Lamb et al, 2005). However, restoration approaches that diminish the provision of ecosystem goods and services that people gain from such degraded land, e.g. through grazing, will not receive lasting support (Maginnis and Jackson, 2005). Ignoring the immediate needs of local populations and replacing their use of land with tree plantations that could not maintain their livelihoods has led to the failure of many such projects in the past (see also Chapter 6). This type of situation highlights the need to design plantations and landscapes containing plantations to accommodate the needs of local people for the range of ecosystem services they depend on. In many regions, plantations have already been effectively used to restore forest ecosystems (Parrotta et al, 1997) and the economic, social or environmental services they provide (Chokkalingam et al, 2006; de Jong, 2010). There is growing evidence that plantations can effectively assume the provision of several ecosystem services, such as maintaining water and nutrient cycles, soil protection and the provision of habitat for biodiversity (see Chapters 3, 4 and 5 in this book). In addition, the focus on ecosystem services in restoration efforts does not contradict but may complement socio-economic objectives such as poverty alleviation (Lamb et al, 2005; Mansourian et al, 2005).

Owing to their very efficient system of wood production, tree plantations are indispensable. The role of plantations has been ambivalent in the past, and therefore there has been much controversy about plantations (Cossalter and Pye-Smith, 2003; Kanowski and Murray, 2008). To improve the role of plantations and to reduce the conflict about their use, new approaches for plantation forestry, including an orientation towards ecosystem services, are required.

Different international processes have aimed to define some principles to guide the future development and management of plantation forests to ensure that they meet not only global but also regional and local demands for all the goods and services they can offer (Chapter 7; Kanowski and Murray, 2008). One example of such principles can be seen in the Voluntary Guidelines for the Responsible Management of Planted Forests (Box 8.1).

As can be seen from Box 8.1, the maintenance or enhancement of ecosystem goods and services is central to all but the institutional principles. In addition, or more explicitly, intelligent solutions for plantations should feature ecological restoration, aim to achieve optimum productivity of multiple products and services rather than maximum productivity of one product and

<hr>

Box 8.1 *FAO principles for the responsible management of planted forests*

Institutional principles
1 Good governance
2 Integrated decision-making and multi-stakeholder approaches
3 Effective organizational capacity

Economic principles
4 Recognition of the value of goods and services
5 Enabling environment for investment
6 Recognition of the role of the market

Social and cultural principles
7 Recognition of social and cultural values
8 Maintenance of social and cultural services

Environmental principles
9 Maintenance and conservation of environmental services
10 Conservation of biological diversity
11 Maintenance of forest health and productivity

Landscape approach principles
12 Management of landscapes for social, economic and environmental benefits

Source: FAO, 2007

<hr>

should be characterized by closed cycles of nutrients and energy conservation. These requirements lead inevitably to knowledge-intensive systems, which is captured in the institutional principles of the FAO Voluntary Guidelines (Box 8.1). The plantations must also be economically profitable from the perspective of investors or owners, and they should also provide economic benefits to local and regional economies.

Where plantation planning and management is following the above principles, plantations can make substantial contributions to the delivery of ecosystems services, many of which are becoming more rather than less important (Kanowski and Murray, 2008).

Plantations as designer ecosystems

Forest plantations are in most situations artificial ecosystems, designed to be simple so that they can be easily and efficiently managed for the purpose of wood production. However, artificial ecosystems must not be simple or have a narrow focus. Recently, designer ecosystems have been proposed to create well-functioning communities of organisms that optimize the ecological services available from coupled natural–human ecosystems (Palmer et al, 2004). Here, the design may not just be concerned with the choice of tree species and the arrangement of trees, but with the creation of entire ecological communities and functioning landscapes to meet specific services, and the use of complex biotic interactions becomes the key technology (Shiyomi and Koizumi, 2001; Kirschenmann, 2007). Such designed plantation ecosystems are not modelled after historical references of ecosystem structure and function for a given location, as is typically the case in ecosystem restoration efforts. Instead, such systems may be designed to mitigate unfavourable conditions by means of novel mixtures of native and non-native species with particular traits that favour specific ecosystem functions (Palmer et al, 2004). They are thus not a substitute for natural systems, but in our highly modified world, they can take over functions of natural systems or ease the pressure on natural systems. In addition, these designed systems can take account of changing environmental conditions in the future, if the knowledge exists, how well the different species and communities will cope with or perform under the new conditions that may be brought about by climate change, species invasions or other forces shaping ecosystems (van der Meer at al, 2002; Seastedt et al, 2008). In more complex settings, plantations may take over the functions of completely different ecosystems such as wetlands or of technical solutions such as sewage treatment plants. For example, in an increasingly urbanized world, safe and environmentally sound ways to treat and dispose of effluent or biosolids produced by industries and urban centres may include the use of irrigated tree plantations (e.g. Hopmans et al, 1990; Myers et al, 1999; Börjesson and Berndes, 2006). In addition to being a safe and low-cost solution to the cleaning of wastewater and recycling of nutrients, these plantations can provide energy, timber and even amenity values. The combination of these functions is likely to be particularly promising in peri-urban areas of arid and semi-arid regions, where a large proportion of available water is appropriated by humans and where the increasing demand for firewood has led to the degradation and depletion of natural forest and woodland systems.

Plantations composed of salt-tolerant species and placed in specific locations in the landscape can assume important ecosystem functions, such as the maintenance of hydrological balance, to uphold the viability of agricultural landscapes in many dry regions of the world. Here, elevated saline groundwater tables, which may be the consequence of irrigation or of clearing deep-rooted perennial vegetation for pastures and cropping or of irrigation, threaten the continuation of agricultural land use (Johnson et al, 2009). Planting of trees to reduce the recharge of groundwater and to increase the discharge from groundwater can help to lower the water table or stop it from rising further to the surface (Chapter 4 in this book; Nambiar and Ferguson, 2005). Depending on the salt concentration of the groundwater and the height of the water table, different types of trees (or shrubs) may be most suitable for this purpose, in many cases non-native species (e.g. Mahmood et al, 2001).

These are just a few examples, where the main purpose of plantations may be on specific ecosystem services, and wood production assumes a secondary role. The conditions under which such purposefully designed plantations are established may be adverse or sub-optimal for tree growing, so that the costs for establishment and maintenance may not be recouped through the production of timber. Where the plantation owner is not also the direct beneficiary of the ecosystem service(s) provided, mechanisms such as payment for ecosystem services (PES) must be developed to reward plantation owners for providing these services. PES is often defined as voluntary, conditional transactions between the buyer and the seller for well-defined environmental services or corresponding land-use proxies (Wunder, 2005). A comparative study carried out by Wunder et al (2008) analysed 14 PES or PES-like programmes in developing and developed countries. The vast majority of the programmes studied were related to management or protection of watersheds through natural forests. Only three of the cases studied included forest and tree plantations (afforestation, reforestation, agroforestry) as eligible activity of the programme, indicating that so far plantations have not been used much within such schemes.

Payments for ecosystem services may be in the form of carbon credits and other credits for specific ecosystem services, e.g. salinity credits. However, where plantations on marginal land for tree growing are required over large areas to restore landscape or watershed functioning, the amount of financial assistance required for tree growers to break even can quickly reach very large sums. For example, Ferguson (2005) calculated for the Murray-Darling Basin in Australia that for eucalypt or pine plantations on land of lower productivity (mean annual increment of 15m/ha/yr) up to AU$2000–3000/ha respectively may be required in the form of salinity credits to achieve a net present value of zero, after accounting for revenues from timber and carbon credits.

One of the important challenges to facilitate schemes such as PES or other reward systems and to facilitate sound landscape planning is to develop ways to measure the ecological functions of plantations (Hartley, 2002; Dudley, 2005). If in future, plantation design and management should consider more fully the whole range of ecosystem goods and services, both agreed systems for valuing goods and services (see Chapter 2) and a good understanding about the compatibility, and possible trade-offs and synergies among ecosystem goods and services are required.

Synergies and trade-offs between ecosystem goods and services

To enable implementation of plantations according to the FAO principles (Box 8.1), in particular in relation to effective, transparent and integrated land-use planning, the costs and benefits and synergies and trade-offs of the many different options that may be available must be known or assessable. Therefore agreed valuation systems for ecosystem goods and services are required (see Chapter 2). Financial valuations often disregard the importance of social and cultural values, although the latter are important and should still have a place in decision-making (see principle 4 in Box 8.1). In economic terms, all goods and services can be defined as use values and non-use values. Socio-cultural values are usually lower in plantations than in natural forests and the ecological values of plantation forests depend largely upon the condition of the landscape replaced by the plantation. Although valuation systems, as discussed in Chapter 2, have a number of shortcomings, they can help to highlight the synergies, trade-offs and implications of different design and management options for plantations.

There is no form of plantation management, or any other form of natural resource management for that matter, that can provide a maximum of all ecosystems goods and services to all stakeholder groups. Some of the services conflict with each other and it would not be possible to try to maximize wood production, carbon sequestration, conservation of biodiversity, and social and cultural benefits in the same plantation stand (Figure 8.1). However, as formulated in the last of the FAO principles for responsible management of planted forests, new approaches would seek a balance of economic, environmental and social objectives at higher spatial scales. With increasing spatial scale, that is moving from one plantation stand or one property to the watershed or landscape, it becomes increasingly easier to reconcile conflicting or non-complementary objectives of management. In addition, in any landscape setting there will be a range of different interest priorities with regard to natural resource management represented by different stakeholder groups or sections of society (Brown, 2005). Satisfying these interests in different parts of the landscape may reduce conflict. Also, many of the ecosystem services depend on ecosystem processes that operate at different spatial and temporal scales, many of which exceed the scale of traditional management (Christensen et al, 1996), such as plantation stand or block and rotations. To appropriately consider these spatial and temporal dimensions, models are required that permit the analysis of spatial and temporal interactions of different types of land use on the provision of ecosystem services in the landscape. The actual planning of plantations in the landscape would be best based on a decision support framework including steps such as environmental and social impact assessments that draw on this information about cost and benefits as well as trade-offs and synergies associated with the different options. An example for such a decision support framework can be found in Kanowski and Murray (2008).

There are several studies dealing with tools to assess the effect of silvicultural techniques on the maintenance and provision of forest goods and services. For example, Köchli and Brang (2005) modelled the effect of different forest management scenarios on recreational suitability and water- and air-purification potential. They show how such an approach may help to explain how different goods and services are interrelated, and what the trade-offs are of the various stand types. Other studies show that the inclusion of high conservation value areas and biodiversity corridors could help to improve

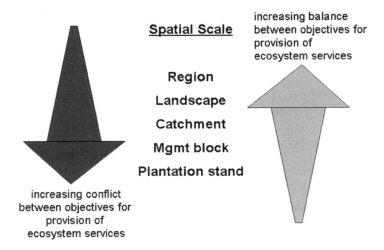

Figure 8.1 *With decreasing scale of management, the conflict between the provision of different ecosystem services or forest values increases. Sustainable solutions aiming at the balanced provision of ecosystem goods and services can only be achieved at higher spatial scales of management*

Source: adapted from Bauhus, 1999

biodiversity levels in plantation areas without affecting the production function (e.g. Barlow et al, 2007; Cyranoski, 2007).

Table 8.1 is an attempt to illustrate these trade-offs between a range of ecosystem goods and services related to the management options at the spatial scale of plantation stands and landscapes. Here, the management encompasses different silvicultural options, which have already been mentioned in Chapter 5, and landscape-level planning options. In the following, some examples from this table, which reflects the information provided in previous chapters, will be explained.

As can be seen from Table 8.1, most of the measures that benefit biodiversity, impact negatively on plantation productivity, both at the stand as well as the landscape level. However, at the landscape level, it would be important to separate between effects that impact on the production per unit of planted land or on the overall plantation estate including other forms of vegetation or land-use types. For example, maintaining corridors of native vegetation instead of converting them to plantation stands, may reduce productivity at the estate or property level, but there is no likely negative influence on the productivity of the plantation stands. Perhaps these are on average even more productive and more efficient to manage, if the native vegetation is occupying parts of the landscape that are less fertile or difficult to cultivate, such as wet soils, steep slopes or rocky outcrops. The synergies

Table 8.1 *Estimated trade-offs between the effects of certain management options on selected ecosystem goods and services including the provision of biodiversity, carbon sequestration or storage, clean water in sufficient quantity, and provision of non-wood forest products*

Management options	Plantation productivity	Biodiversity	Carbon[#]	Water*	Amenity values
Stand level					
Structural retention	–	+	?	+	+
Use of native species	–	+	(–)	(+)	+
Mixed-species stands	+	+	+	–	+
Long rotations	–	+	?	+	+
Thinning	0	+	–	+	(+)
Site preparation	+	–	?	?	–
Herbicides and fertilizer	+	–	?	–	–
Landscape level					
Riparian buffers (of native vegetation)	–	+	(+)	+	+
Retaining patches of native vegetation	–	+	(+)	+	+
Connectivity between plantations and native forests	0	+	0	0	+
Maintaining landscape heterogeneity (different land-use types, special places, etc.)	–	+	?	+	+

Regarding the influence of plantations to reduce atmospheric CO_2, sequestration and storage need to be separated. Sequestration, the uptake of carbon into vegetation and its partial transfer into the soil pool, removes CO_2 from the atmosphere. This process is tightly coupled with plantation productivity. Storage of C in vegetation and soils simply prevents C from being released to the atmosphere as CO_2. Forest systems such as plantations may have a high sequestration potential but little storage, whereas the opposite situation can be found in old-growth forests. For many of the management options it is difficult to ascertain the effect that they have on atmospheric CO_2 since this depends to a large extent on the fate of the material harvested (see Chapter 3). If the harvested wood is turned into long-lived products, which, at the end of their service live, are used energetically to offset fossil fuel burning, the effect can be very positive. In contrast, if the wood is turned into short-lived products (such as paper), and is not subsequently used as an energy source, the overall effect may be less than if the wood was left in the forest to decay (Profft et al, 2009). Therefore, the ultimate effect of plantations on C cycling and atmospheric CO_2 cannot be assessed within the plantation management system.

* For the purpose of this assessment, management options were considered to have a positive effect on water services, if they contribute to cleaner water and more groundwater recharge from the land (but see discussion on salinity, Chapter 4). The effect is considered negative, if planning or management options lead to water pollution and reduced groundwater recharge from the land carrying plantations.

Note: + = positive effects, – = negative effects, 0 = neutral effect, ? = unknown or uncertain effects, brackets indicate that the effect may not be so clearly positive or negative depending on other factors not captured here

and trade-offs between biodiversity and productivity at the landscape level depend largely on the forest policy context. If plantation establishment is directly related to and dependent on the area of native forests set aside for conservation, there can be strong synergistic effects (Paquette and Messier, 2010). In addition, there are also options to increase biodiversity values that have little or no additional costs (see Chapter 5).

Most measures that are suited to improving biodiversity in plantation landscapes also have positive effects on amenity values. Owing to their orderliness and their uniformity in shape, structure and composition, plantations can have a dramatic impact on amenity and recreational values (Evans, 2009). Artificial boundaries, strong contrasts and sharp edges between stands and large clear-felled areas or other land-use types and the monotony of landscapes dominated by single species are important aspects of how the public perceives planted forests, although perception may also be influenced by the designation of the forested landscape (Anderson, 1981) and the history of afforestation and forest use (Ní Dhubháin et al, 2009). However, the preservation of patches of native vegetation and the maintenance of landscape heterogeneity as well as the creation of stand structural diversity, are likely to benefit the cultural services of planted forests.

In general, there are mostly synergies between the supporting ecosystem services such as the maintenance of soil resources, water and nutrient cycles, and biodiversity. However, most of the above-listed management options that have a direct or indirect negative effect on plantation productivity, have a positive influence on water services and vice versa (e.g. Vertessy et al, 1996; Jackson et al, 2005). However, this does not apply universally to all other – including undesirable and unintended – negative effects on productivity, as for example through soil compaction or erosion. Here, the focus is on effects on the physiological activity and transpirational demand of planted forests. Where measures such as structural retention or longer rotations result in fewer young and vigorously transpiring trees on site, the water demand of plantations will also decline. Less productive native species are likely to consume less water, than more productive exotics such as eucalypts, which have been criticized for their high water demand (Calder, 2002). Long rotations are likely to reduce the average plantation productivity, if the rotations are extended substantially beyond the culmination of mean annual increment. Slower-growing, older plantations will have a lower transpirational demand (Vertessy et al, 1996) and less frequent disturbances will also lead to overall improved water quality (Croke et al, 2001). Reduced transpiration and interception following thinning will increase water yield from plantations (e.g. Breda et al, 1995), albeit only for a limited period of time (e.g. Lane and Mackay, 2001), while thinning is unlikely to have negative impacts on water quality, except through the use of forest roads and extraction tracks. In contrast, typical plantation management practices such as site preparation and the use of herbicides and fertilizer have the potential to reduce water quality through the disturbance of soil, removal of protective soil cover and addition of nutrients, which may not be taken up by the vegetation (e.g. Malmer, 1996). However, these effects can be minimized through adherence to appropriate codes of forest practice.

Options to improve water services from plantations at the landscape scale are largely related to the percentage area under plantation, the specific location of plantations and the protection of soils and waterways (see Chapter 4). Here, the services can be optimized through policy settings, certification requirements or sound landscape planning (see Chapter 7). However, stand- and landscape-level management options to improve water services of plantations are unlikely to conflict with the provision of other ecosystem goods and services besides the production function (e.g. Wang et al, 2009).

Synergies and trade-offs with other ecosystem goods and services are most difficult to identify for the carbon sequestration or storage function of plantations, because the influence of the above-considered management options depends on many other factors, such as the fate of the harvested wood. However, most options that increase productivity at the stand level, are likely to also increase the sequestration of carbon, unless these increases are at the expense of soil stored C, which may be the case for site preparation. While the effects of site preparation on C storage are likely to be negative in the short term, in particular in relation to soil C (Paul et al, 2002), increased productivity in the long term may offset these reductions. While C sequestration may be reduced through longer rotations owing to reduced productivity, it is likely that more C is stored on site.

At the landscape scale, interspersing plantations with buffer strips and reserves of native vegetation creates patches with more long-term C storage than in the plantation stands, although the C sequestration in these patches may be less than in the highly productive plantations. Table 8.1 illustrates that focusing plantation management on mitigating climate change through C sequestration and possibly replacing fossil fuels through bioenergy, may have serious implications on many other ecosystem services, in particular if perverse incentives are provided for some short-term goals (e.g. Danielsen et al, 2008).

At the stand scale, the production of wood and fibre has limited synergies with the other ecosystem services listed here, except for C sequestration. Typical measures to increase plantation productivity such as site preparation and the use of fertilizers and pesticides have no direct beneficial effects on the other ecosystems services. However, Table 8.1 provides a very general and simple perspective on trade-offs and conflicts between services, which are often specific to the site and context. The concept shows that it is necessary to identify how different ecosystem goods and services may be differently affected to optimize their provision at various scales. The table also shows that synergies are easier to achieve and trade-offs easier to avoid at the landscape scale when compared to the stand scale. However, higher provision of other ecosystem services often comes at the expense of plantation productivity. Therefore one of the important challenges is to devise mechanisms to reward landowners for these other plantation functions that may conflict with the production of conventional plantation products. However, only in very few cases have ecosystem goods and services been quantified for a particular plantation setting (e.g. Nambiar and Ferguson, 2005; Barlow et al, 2007). And few studies have aimed to explore what the optimal spatial aggregation of the various stand types should be to deliver the best mix of goods and services. To achieve this, several studies suggest a combination of forest growth models, geographic information systems (GIS) and indices of goods and services to support the development of land-use visions and forest planning policies on a regional scale (e.g. Köchli and Brang, 2005).

So far many challenges remain with incorporating and applying the ecosystem goods and services approach in actual design, planning and management at the landscape level (de Groot et al, in press). Only a few ad hoc attempts have been made to use spatial planning as a tool to improve the overall yield of ecosystem goods and services and to find the appropriate level of trade-offs and synergies at the landscape level. For instance Van Eupen et al (2007) used scenario-dependent maps, which indicate habitat suitability of landscapes for certain flagship species. When integrated with planning for the other most important provisioning, regulating, cultural and supporting goods and services, this could be used in new approaches for plantation planning, for example in the restoration of degraded forest landscapes in tropical areas to address both sustainable use of biodiversity as well as the alleviation of rural poverty (e.g. Lamb et al, 2005).

Unfortunately, cost-effective monitoring approaches have not been developed for many of the ecosystem goods and services, which would demonstrate and quantify the many benefits and impacts of plantations. Therefore, and for many other reasons, the identification, evaluation and negotiating of trade-offs is rarely done in the process of plantation planning in the landscape.

Particularly difficult is the assessment of temporal trade-offs between short-term benefits and the long-term capacity of ecosystems to provide services to future generations (Chapin, 2009). Most measures that maintain the productive capacity of plantation systems, in particular the maintenance of soil resources, should also maintain the natural capital in the long term. However, here the interactions with other systems such as fresh water or the atmosphere also have to be considered. For example, one would have to question the usefulness of sequestering more atmospheric CO_2 now by increasing plantation productivity through high use of N fertilizers which may lead to denitrification and increases in atmospheric concentrations of N_2O. The latter gas has, owing to its longevity of around 120 years, an estimated global warming potential (based on a 100-year period) that is 310 times as high as that of CO_2 (IPCC, 1996). Temporal trade-offs occur in particular for those ecosystem services that cannot be restored once they are lost. These include loss of species or fossil groundwater, which are at least as valuable in the future as they are now (Heal, 2000).

How plantations may help to solve some of the urgent global issues

The Millennium Ecosystem Assessment (2005) has demonstrated that the well-being of mankind depends on the maintenance or improvement of ecosystem services. It is increasingly recognized that plantations can make major contributions to the provisioning of ecosystem goods and services (Paquette and Messier, 2010), but they can also have negative impacts. Therefore, we ask here, which contributions plantations can make, through the ecosystem goods and services they provide, to solve some of the pressing global problems that are currently faced by mankind. Tree plantations already play a major role in the provision of timber and fibre such that around 3 per cent of the forest area, which is in productive plantations, provide approximately one-third of the industrial roundwood worldwide (Chapter 1). This is a success story and shall not be further discussed here. Below we discuss the different roles that plantations can play in solving the biodiversity and energy crisis, in mitigating climate change, and in reducing poverty.

Biodiversity

In Chapter 5 of this book, it has been shown that plantations can be managed and planned in ways that improve their habitat value over that of conventionally managed plantations. Beneficial effects of plantations can be expected, in particular, where these are used to restore degraded land or replace agricultural systems, and where they increase connectivity in the landscape. In addition, many options exist at the stand level to enhance the provision of habitat through increasing structural complexity, species mixtures, prolonged rotations and alternative methods of site preparation. However, the greatest benefit from plantations for biodiversity is likely to stem from the highly efficient wood production system they present.

The high efficiency of wood production that is possible in tree plantations may be the foundation for setting aside more areas of native forests in conservation reserves (e.g. Côté et al, 2009) or to reduce the management intensity in native or semi-natural forests. As was pointed out in Chapter 1, if all industrial wood came from effectively managed planted forests, only some 73 million ha, i.e. only less than 2 per cent of the world's forest area would be enough to satisfy the current global need for industrial wood (Seppälä, 2007). Given the current extent of plantations, these figures also indicate that a substantial proportion is not as effectively managed as possible and that rather than expanding their area, it may be more promising to invest in improvements in their management. In any case, fast-growing, industrial tree plantations may be one of the most effective approaches for the conservation or improved management of other forest areas valued for their biodiversity, beauty and other things. In return, the reservation of some or the majority of native forests from wood production will act as a driver for the establishment of further plantations, as has been the case for example in Australia (URS Forestry, 2007).

Energy

Energy from biomass traditionally plays a very important role in countries with developing economies. Recently, the global demand for bioenergy has been accelerating because of the depletion of fossil fuels and the CO_2 reduction targets of most countries. This is putting further pressure on forested areas. For instance, over the last five years there has been a rapid development of new bio-fuel plantations in Malaysia and Indonesia (mainly oil palm) and Brazil (mainly soybean, maize) (e.g. Mantel et al, 2007; Reinders and Huijbregts, 2008; Fargione et al, 2008). This has often gone at the expense of natural forest areas. But also forests themselves, either planted or natural, are increasingly being used for bioenergy. In Europe, fast-growing short-rotation willow and poplar plantations are being used for bioenergy (chips) (e.g. Börjesson and Berndes, 2006), and also in tropical countries fast-growing species such as *Acacia mangium* and *Eucalyptus* sp. are increasingly being used for bioenergy production (Berndes et al, 2003). This is likely to continue in the near future: Buerkert and Schlecht (2009) estimate that in 2050 there may be 500Mha of new fuelwood plantation established.

Energy is widely recognized as a crucial factor for health, education, and economic development. Renewable forms of energy provide countries with developing economies with the opportunity to reduce their dependency on costly fossil fuels, and they are often the most cost-efficient way of improving energy services for rural areas, which are often not connected to the electrical grid (Shukla et al, 2004). Therefore, biomass burning is by far the most widely used form of energy in many African countries, and will remain so for the foreseeable future (Karekezi and Kithyoma, 2002). Biomass is the only form of energy of relevance to most households, especially the poor. The present pattern of consumption of wood, however, is clearly not sustainable, and will soon result in irrecoverable damage to life-sustaining ecosystems. Here, decentralized and community-based small-scale electrification schemes based on biomass from plantations can deliver affordable and sustainable power in rural areas, without which development beyond a certain level is impossible. Obviously, the fitting type of plantation required for this purpose would also be community-based, even though they may require larger areas.

New developments in bioenergy production are opening opportunities for forest plantations as a sustainable source of biomass, but they are also posing several challenges. Under the increased demand for energy crops, prices can be expected to rise and this will enable good management practices including sound ecological, economic and social production (Rootzén et al, 2010). When wood from short-rotation plantations is used for energy purposes (replacing fossil fuels) these plantations also have a high climate change mitigation potential (Rootzén et al, 2010; Chapter 3). These fast-growing plantations also have the potential to combine a range of services (see Table 8.1); however, the interactions between fuelwood plantations and other land uses such as agriculture and nature conservation, and with functions such as carbon sequestration have been insufficiently analysed (e.g. Berndes et al, 2003).

Climate change mitigation

Owing to their capacity to sequester atmospheric CO_2, there has recently been much interest in using plantations in climate change mitigation strategies. In Chapter 3, five strategies for the contribution of the forestry sector have been identified: (1) increasing the forest area through afforestation; (2) increasing the carbon stored in existing forests; (3) protecting existing carbon stocks from release into the atmosphere; (4) increasing the carbon stored in products (yielding also indirect greenhouse gas mitigation through material substitution); and (5) substituting fossil fuels with bioenergy derived from forest biomass and wood (see above).

The greatest realistic potential for plantations appears to be in the first and in the fifth options. However, carbon sequestration in the forest-based sector is largely a non-permanent strategy. The sequestration phase is finite. In plantations it may last only for some decades and then the gained carbon stocks would need to be protected to keep carbon withdrawn from the atmosphere. While this is not possible within a single plantation stand, it may well possible over larger temporal and spatial scales, if plantations are maintained and not converted back into agricultural land use. Sequestration therefore always needs to be protected by safeguarding measures to make mitigation strategies effective.

Extending the length of rotations in existing forests is currently unlikely to receive the financial reward required to compensate for not harvesting trees at their economical maturity. Increasing the proportion of wood supply from fast-growing and sustainably managed plantations to permit a reduction of harvesting and the setting aside of native forests for conservation purposes, as we have discussed above in the context of biodiversity values, may also be a substantial contribution of plantations to the protection of carbon stocks in these forests, which may be old-growth forests with high C density. Increasing the use of long-lasting wood products and improving recycling rates has the potential to reduce the pressure on forest resources; however, this is a broader issue that is not exclusive to plantation products.

Plantations, however, may not only play a role in mitigation strategies to climate change. They may also be strongly affected by climate change with ramifications for the goods and services provided by them. Adapting plantations to climate change is another challenge. Compared to native or semi-natural forest with slow-growing, long-lived trees, this task may be much easier in fast-growing plantations, where changes of species, provenances or clones to adapt to climate-induced changes in site conditions can be accommodated frequently between rotations. However, the introduction of new species may also require the adaptation of silvicultural practices and the processing chain, which may be rather difficult and costly.

Poverty reduction and the social effects of plantations

The immense productivity of plantations and attractive markets for plantation products have fuelled expectations in governments and experts of positive economic effects at regional and national level, in particular in the context of non- or under-developed rural regions in the tropics and subtropics (UN, 2002). Beyond the demand on plantation products, these expectations are one of the reasons, respectively justifications, for the still significant subsidies and numerous plantation programmes financed by governments and international organizations (Cossalter and Pye-Smith, 2003). However, with the expansion of plantation forestry and rising criticism, mostly about negative environmental effects, the public has also become more and more sensitive to the social dimensions of plantations (Chapter 6; Carrere, 1998). While some studies highlight the positive effects of plantations on employment and income generation, in particular in outgrower schemes (e.g. Mayers and Vermeulen, 2002), as well as the indirect environmental benefits from reforestation programmes, other studies have drawn a more critical picture in light of land conflicts, limited employment opportunities and environmental damage caused by plantations (Hoch et al, 2009). The social conflicts have been particularly apparent, where intensively managed forest plantations have been implemented (Kanowski and Murray, 2008). Whilst the social balance in landscapes influenced by plantations depends on the specific context and design, it seems possible to attribute the social performance to a larger degree to the functioning of interactions between actor groups (see Chapter 6).

To better understand the social dimensions of plantations, it is useful to distinguish between the scale and owners of plantations: (a) large companies and governments that own or manage plantation resources from tens of thousands to hundreds of thousands of hectares; (b) independent private landowners who typically manage between hundreds to thousands of hectares; and (c) smallholders with resources typically in the range of a few hectares or even less; see also Kanowski and Murray (2008).

Owing to their capacities regarding capital and know-how, companies, governments and other actors with sufficient capital are predestined to initiate large-scale forestry plantations. The economic benefit of these types of plantations is often expressed as the return to investors, which has been in the range of 3–11 per cent internal rate of return for short-rotation pulpwood and 1–7 per cent for longer-rotation solid-wood plantations (RISI, 2007). The most important social benefit of these types of plantations at a local level is through employment. However, in consideration of the immense demand on land, the employment opportunities generated by the tree growing in plantations are rather limited, if compared to other more labour-intensive land uses such as family agriculture (Cossalter and Pye-Smith, 2003; Schirmer and Tonts, 2003). Against this background, the social balance of these type of plantations depends mainly on the attractiveness of alternative land-use options given by contextual parameters such as population density, land-use history and, related to this, soil fertility. Generally, the social impacts of large-scale industrial plantations, due to their immense requirement for land, tend to be more negative in highly populated areas with fertile soils. Most of the economic benefits of plantations are actually associated with the processing industries reliant on the plantation products (Kanowski, 2005) and hence the employment opportunities. These processing industries have often not been established in the same areas as the plantations. However, in some cases even the local commercialization of by-products such as quality timber, fuelwood or charcoal may provide significant local benefits. In addition, large-scale industrial plantation initiatives may catalyse and provide the basis for the participation of the other two types of plantation owners – (b) and (c) from above – in the business of tree growing.

For example in outgrower schemes, farmers and smallholders are actively engaged in the production process and may directly benefit from attractive income opportunities. However, despite some outstanding positive experiences (e. g. Mayers and Vermeulen, 2002), the competitive disadvantages of locals in negotiating with companies bears the risk of unfair contracts (Desmond and Race, 2000).

Many national and international development organizations have also been promoting plantations owned and managed by smallholders to generate income and to achieve environmental goals. While there have been some very positive local experiences, the success of such initiatives has been rather modest (Chapter 6; Hoch et al, 2009). The success and problems of this approach have been discussed in detail in Chapter 6. The problems typically associated with this approach are:

- the establishment of plantations on degraded soil resulting in poor tree growth;
- the technologies, capital and security of tenure required are often not compatible with smallholders' realities;
- the dependence on continuous external support not only for the establishment but also for the maintenance and, most important, the commercialization of plantation products.

As has been shown in Chapter 6, it is often overlooked by development practitioners that smallholders also grow trees on their own initiative, often as single trees within agricultural crops or in homegardens and in small plantations. Often these plantings focus on the production of NTFPs and other ecosystem services, which provide more immediate and regular benefits than wood and timber.

Given these different situations and contexts, the role of plantations in poverty reduction has been ambivalent. The conditions and prerequisites for the positive social effects of plantations at different scales have been outlined in Chapter 6 and other recent publications (e.g. Kanowski and Murray, 2008). Positive effects will only be realized if all the costs and benefits of the development of plantation landscapes are being assessed and where the legitimate interests of all actors are respected. Where this is not the case, plantations will not only fail to contribute to poverty reduction, they are also unlikely to provide ecosystem goods and services to those who are most strongly dependent on them.

Conclusions

The increasing competition for land in an increasingly crowded world may result in a shift in management focus of existing and future plantations, from single or few goals such as efficient wood production, to a range of goals. In the past, the dominant interests in societies often accorded little formal value to the non-timber ecosystem goods and services of plantation forests. Changing societal values, more progressive thinking about the potential roles of plantations, and the emergence of new forms of environmental governance regimes have created policy environments which are much more enabling of the provision of ecosystem goods and services from plantation forests.

In the different chapters of this book it has been documented that there is now a large body of knowledge and experience showing that appropriately planned, designed and managed plantation forests can deliver a range of ecosystem goods and services, at both landscape and stand scales. This book has also shown that the benefits and impacts of plantations and their trade-offs are highly context specific. Therefore, the range of impacts and benefits associated with plantation forests, which typically differ from those of other landscape components, need to be assessed, agreed and managed in a landscape context. This body of knowledge about the impacts of plantations and their contributions to ecosystem goods and services provides the foundation for developing governance regimes that are consistent with the principles of sustainable forest management. In accordance with these widely agreed principles, societies have the right to expect that plantation forests will deliver more benefits than costs. However, how the different ecosystems goods and services are traded off and to what extent different interest groups share the benefits will remain, ultimately, a value judgement. It is clear, however, that enhancing the provision of ecosystem goods and services from plantation forests is an important element of realizing the benefits of this increasingly important form of forestry.

References

Anderson, L. M. (1981) 'Land use designations affect perception of scenic beauty in forest landscapes', *Forest Science*, vol 27, pp392–400

Barlow, J., Gardner, T. A., Araujo, I. S., Avila-Pires, T. C., Bonaldo, A. B., Costa, J. E., Esposito, M. C., Ferreira, L. V., Hawes, J., Hernandez, M. M., Hoogmoed, M. S., Leite, R. N., Lo-Man-Hung, N. F., Malcolm, J. R., Martins, M. B., Mestre, L. A. M., Miranda-Santos, R., Nunes-Gutjahr, A. L., Overal, W. L., Parry, L., Peters, S. L., Ribeiro-Junior, M. A., da Silva, M. N. F., Motta, C. D. and Peres, C. A. (2007) 'Quantifying the biodiversity value of tropical primary, secondary, and plantation forests', *Proceedings of the National Academy of Sciences of the United States of America*, vol 104, pp18555–18560

Barr, C. (2002) *Banking on Sustainability: Structural Adjustment and Forestry Reform in Post-Suharto Indonesia*, CIFOR and WWF, Bogor, Indonesia

Bauhus, J. (1999) 'Silvicultural practices in Australian native State forests – An introduction', *Australian Forestry*, vol 62, pp217–222

Berndes, G., Hoogwijk, M. and van den Broek, R. (2003) 'The contribution of biomass in the future global energy supply: A review of 17 studies', *Biomass and Bioenergy*, vol 25, pp1–28

Börjesson, P. and Berndes, G. (2006) 'The prospects for willow plantations for wastewater treatment in Sweden', *Biomass and Bioenergy*, vol 30, pp428–438

Breda, N., Granier, A. and Aussenac, G. (1995) Effects of thinning on soil and tree water relations, transpiration and growth in an oak forest (*Quercus petraea* (Matt) Liebl), *Tree Physiology*, vol 15, pp295–306

Brown, K. (2005) 'Addressing trade-offs in forest landscape restoration', in S. Mansourian, D. Vallauri, and N. Dudley (eds) *Forest Restoration in Landscapes: Beyond Planting Trees*, Springer Verlag, New York, pp59–62

Buerkert, A. and Schlecht, E. (2009) 'The biofuel debate – status quo and research needs to meet multiple goals of food, fuel and ecosystem services in the tropics and subtropics', *Journal of Agriculture and Rural Development in the Tropics and Subtropics*, vol 110, pp1–8

Calder, I. R. (2002) 'Eucalyptus, water and the environment', in J. J. W. Coppen (ed.) *Eucalyptus: The Genus Eucalyptus*, Taylor and Francis, London, pp36–51

Carrere, R. (1998) *Ten Replies to Ten Lies*, World Rainforest Movement, www.wrm.org.uy/plantations/material/lies.html, accessed 5 March 2010

Chapin, F. S., III (2009) 'Managing ecosystems sustainably: The key role of resilience', in F. S. Chapin, III, G. P. Kofinas and C. Folke (eds) *Principles of Ecosystem Stewardship: Resilience Based Natural Resource Management in a Changing World*, Springer Verlag, New York, pp29–53

Chokkalingam, U., Carandang, A. P., Pulhin, J. M., Lasco, R. D., Peras, R. J. J. and Toma, T. (2006) *One Century of Forest Rehabilitation in the Philippines: Approaches, Outcomes and Lessons*, CIFOR, Bogor, Indonesia

Christensen, N. L., Bartuska, A. M., Brown, J. H., Carpenter, S., D'Antonio, C., Francis, R., Franklin, J. F., MacMahon, J. A., Noss, R. F., Parsons, D. J., Peterson, C. H., Turner, M. G. and Woodmansee, R. G. (1996) 'The report of the Ecological Society of America Committee on the scientific basis for ecosystem management', *Ecological Applications*, vol 6, pp665–691

Cossalter, C. and Pye-Smith, C. (2003) 'Fast-wood forestry: myths and realities', *Forest Perspectives*, CIFOR, Jakarta

Côté, P., Tittler, R., Messier, C., Kneeshaw, D. D., Fall, A. and Fortin, M.-J. (2009) 'Comparing different forest zoning options for landscape-scale management of the boreal forest: Possible benefits of the TRIAD', *Forest Ecology and Management*, vol 259, pp418–427

Croke, J., Hairsine, P. and Fogarty, P. (2001) 'Soil recovery from track construction and harvesting changes in surface infiltration, erosion and delivery rates with time', *Forest Ecology and Management*, vol 143, pp3–12

Cyranoski, D. (2007) 'Logging: The new conservation', *Nature*, vol 446, pp608–610

Danielsen, F., Beukema, H., Burgess, N. D., Parish, F., Brühl, C. A., Donald, P. F., Murdiyarso, D., Phalan, B., Reijnders, L., Struebig, M. and Fitzherbert, E. B. (2008) 'Biofuel plantations on forested lands: Double jeopardy for biodiversity and climate', *Conservation Biology*, vol 23, pp348–358

de Groot, R. S., Alkemade, R., Braat, L., Hein, L. and Willemen, L. (2010) 'Challenges in integrating the concept of ecosystem services and values in landscape planning, management and decision making', *Ecological Complexity* (in press)

de Jong, W. (2010) 'Forest rehabilitation and its implication for forest transition theory', *Biotropica*, vol 42, pp3–9

Desmond, H. and Race, D. (2000) *Global Survey and Analytical Framework for*

Forestry Out-grower Arrangements, FAO, Rome

Dudley, N. (2005) 'Best practice for industrial plantations', in S. Mansourian, D. Vallauri and N. Dudley (eds) *Forest Restoration in Landscapes: Beyond Planting Trees*, Springer Verlag, New York, pp392–397

Evans, J. (2009) 'The history of tree planting and planted forests', in J. Evans (ed.) *Planted Forests: Uses, Impacts and Sustainability*, CAB International and FAO, Wallingford and Rome, pp5–22

Evans, J. and Turnbull, J. (2004) '*Plantation Forestry in the Tropics*', 3rd edn, Oxford University Press, Oxford

FAO (2007) *Voluntary Guidelines: Responsible Management of Planted Forests*, www.fao.org/forestry/plantedforestsguide/en/, accessed 21 January 2010

FAO (2009) 'How to feed the world in 2050', High Level Expert Forum, The Food and Agricultural Organization, Rome, 12–15 October, www.fao.org/wsfs/forum2050/wsfs-background-documents/issues-briefs/en/, accessed 20 January 2010

Ferguson, I. (2005) 'Economic and policy implications', in S. Nambiar and I. Ferguson (eds) '*New Forests: Wood Production and Environmental Services*', CSIRO Publishing, Collingwood, Victoria, pp209–235

Heal, G. (2000) *Nature and the Marketplace: Capturing the Value of Ecosystem Services*, Island Press, Washington, DC

Hoch, L., Pokorny, B. and de Jong, W. (2009) 'How successful is tree growing for smallholders in the Amazon?', *International Forestry Review*, vol 11, pp299–310

Hopmans, P., Stewart, H. T. L., Flinn, D. W. and Hillman, T. J. (1990) 'Growth, biomass production and nutrient accumulation by seven tree species irrigated with municipal effluent at Wodonga, Australia', *Forest Ecology and Management*, vol 30, pp203–211

Hunter, I. R., Hobley, M. and Samle, P. (1998) 'Afforestation of degraded land – pyrrhic victory over economic, social and ecological reality?', *Ecological Engineering*, vol 10, pp97–106

Imhoff, M. L., Bounoua, L., Ricketts, T., Loucks, C., Harris, R. and Lawrence, W. (2004) 'Global patterns in human consumption of net primary production', *Nature*, vol 429, pp870–873

IPCC (1996) *Second Assessment Report – Report of Working Group I – the Science of Climate Change, with a Summary for Policymakers* (SPM), Houghton, J. T., Meira Filho, L. G., Callender, B. A., Harris, N., Kattenberg, A. and Maskell, K. (eds) Cambridge University Press, Cambridge

Jackson, R. B., Jobbagy, E., Avissar, R., Baidya Roy, S., Barrett, D. J., Cook, C. W., Farley, K. A., le Maitre, D. C., McCarl, B. A. and Murray, B. C. (2005) 'Trading water for carbon with biological sequestration', *Science*, vol 310, pp1944–1947

Johansson, D. J. A. and Azar, C. (2007) 'A scenario based analysis of land competition between food and bioenergy production in the US', *Climatic Change*, vol 82, pp267–291

Johnson, I., Coburn, R. and Barton, C. (2009) 'Tree planting to control salinity', Primefact 983, Department of Primary Industries, New South Wales, www.dpi.nsw.gov.au/primefacts

Kanowski, P. (2005) *Intensively Managed Planted Forests*, www.theforestsdialogue.org,

Kanowski, P. and Murray, H. (2008) *Intensively Managed Planted Forests: Toward Best Practice*, www.theforestsdialogue.org

Karekezi, S. and Kithyoma, W. (2002) 'Renewable energy strategies for rural Africa: Is a PV-led renewable energy strategy the right approach for providing modern energy to the rural poor of sub-Saharan Africa?', *Energy Policy*, vol 30, pp1071–1086

Kirschenmann, F. L. (2007) 'Potential for a new generation of biodiversity in agroecosystems of the future', *Agronomy Journal*, vol 99, pp373–376

Köchli, D. A. and Brang, P. (2005) 'Simulating effects of forest management on selected public forest goods and services: A case study', *Forest Ecology and Management*, vol 209, pp57–68

Lamb, D. (1998) 'Large-scale ecological restoration of degraded tropical forest lands – the potential role of timber plantations', *Restoration Ecology*, vol 6, pp271–279

Lamb, D., Erskine, P. D. and Parrotta, J. A. (2005) 'Restoration of degraded tropical forest landscapes', *Science*, vol 310, pp1628–1632

Lambin, E. F., Turner, B. L., Geist, H. J., Agbola, S. B., Angelsen, A., Bruce, J. W., Coomes, O. T., Dirzo, R., Fischer, G., Folke, C., George, P. S., Homewood, K., Imbernon, J., Leemans, R., Li, X., Moran, E. F., Mortimore, M., Ramakrishnan, P. S., Richards, J. F., Skånes, H., Steffen, W., Stone, G. D., Svedin, U., Veldkamp, T. A., Vogel, C. and Xu, J. (2001) 'The causes of land-use and land-cover change: Moving beyond the myths', *Global Environmental Change*, vol 11, pp261–269

Lane, P. N. J. and Mackay, S. M. (2001) 'Streamflow response of mixed-species eucalypt forests to patch cutting and thinning treatments', *Forest Ecology and Management*, vol 143, pp131–142

Maginnis, S. and Jackson, W. (2005) 'Balancing restoration and development', *ITTO Tropical Forest Update*, vol 15, pp3–6

Mahmood, K., Morris, J., Collopy, J. and Slavich, P. (2001) 'Groundwater uptake and sustainability of farm plantations on saline sites in Punjab province, Pakistan', *Agricultural Water Management*, vol 48, pp1–20

Malmer, A. (1996) 'Hydrological effects and nutrient losses of forest plantation establishment on tropical rainforest land in Sabah, Malaysia', *Journal of Hydrology*, vol 174, pp129–148

Mansourian, S., Vallauri, D. and Dudley, N. (eds) (2005) 'Forest restoration in landscapes: Beyond planting trees', Springer, New York

Mayers, J. and Vermeulen, S. (2002) *Company–Community Forestry Partnerships: From Raw Deals to Mutual Gains?* International Institute for Environment and Development, London

Metzger, J. O. and Hüttermann, A. (2009) 'Sustainable global energy supply based on lignocellulosic biomass from afforestation of degraded areas', *Naturwissenschaften*, vol 96, pp 279–288

Millenium Ecosystem Assessment (MEA) (2005) *'Ecosystems and Human Well-being, Synthesis'*, Island Press, Washington, DC

Myers, B. J., Bond, W. J., Benyon, R. G., Falkiner, R. A., Polglase, P. J., Smith, C. J., Snow, V. O. and Theiveyanathan, S. (1999) *Sustainable Effluent-Irrigated Plantations – An Australian Guideline*, CSIRO Forestry and Forest Products, Melbourne

Nambiar, S. and Ferguson, I. (eds) (2005) *New Forests: Wood Production and Environmental Services*, CSIRO Publishing, Collingwood, Victoria

Ní Dhubháin, A., Fléchard, M.-C., Moloney, R. and O'Connor, D. (2009) 'Stakeholders' perceptions of forestry in rural areas – Two case studies in Ireland', *Land Use Policy*, vol 26, pp695–703

Otsamo, A. (2001) 'Forest plantations on *Imperata* grasslands in Indonesia – Establishment, silviculture and utilization potential', PhD thesis, University of Helsinki, www.mm.helsinki.fi/mmeko/vitri/studies/theses/otsamothes.pdf, accessed 24 May 2010

Palmer, M., Bernhardt, E., Chornesky, E., Collins, S., Dobson, A., Duke, C., Gold, B., Jacobson, R., Kingsland, S., Kranz, R., Mappin, M., Martinez, M. L., Micheli, F., Morse, J., Pace, M., Pascual, M., Palumbi, S., Reichman, O. J., Simons, A., Townsend, A. and Turner, M. (2004) 'Ecology for a crowded planet', *Science*, vol 304, pp1251–1252

Paquette, A. and Messier, C. (2010) 'The role of plantations in managing the world's forests in the Anthropocene', *Frontiers in Ecology and the Environment*, vol 8, pp27–34

Parrotta, J. A., Turnbull, J. W. and Jones, N. (1997) 'Catalyzing native forest regeneration on degraded tropical lands', *Forest Ecology and Management*, vol 99, pp1–7

Paul, K.I., Polglase, P. J., Nyakuengama, J. G. and Khanna, P. K. (2002) 'Change in soil carbon following afforestation', *Forest Ecology and Management*, vol 168, pp241–257

Postel, S. K., Daily, G. C. and Ehrlich, P. R. (1996) 'Human appropriation of renewable fresh water', *Science*, vol 271, pp785–788

Profft, I., Mund, M., Weber, G.-E., Weller, E. and Schulze, E. D. (2009) 'Forest management and carbon sequestration in wood products', *European Journal of Forest Research*, vol 128, pp399–413

RISI (2007) *The Global Tree Farm and Managed Forest Industry, Multi-Client Study, February 2007*, RISI, Inc., www.risiinfo.com, accessed 24 May 2010

Rootzén, J. M., Berndes, G., Ravindranath, N. H., Somashekar, H. I., Murthy, I. K., Sudha, P. and Ostwald, M. (2010) 'Carbon sequestration versus bioenergy: A case study from South India exploring the relative land-use efficiency of two options for climate change mitigation', *Biomass and Bioenergy*, vol 34, 116–123

Schirmer, J. and Tonts, M. (2003) 'Plantations and sustainable rural communities', *Australian Forestry*, vol 66, pp67–74

Schulze, E. D., Luyssaert, S., Ciais, P., Freibauer, A., Janssens, I. A., Soussana, J. F., Smith, P., Grace, J., Levin, I., Thiruchittampalam, B., Heimann, M., Dolman, A. J., Valentini, R., Bousquet, P., Peylin, P., Peters, W., Rodenbeck, C., Etiope, G., Vuichard, N., Wattenbach, M., Nabuurs, G. J., Poussi, Z., Nieschulze, J. and Gash, J. H. (2009) 'Importance of methane and nitrous oxide for Europe's terrestrial greenhouse-gas balance', *Nature Geoscience*, vol 2, pp842–850

Seastedt, T. R., Hobbs, R. J. and Suding, K. N. (2008) 'Management of novel ecosystems: Are novel approaches required?', *Frontiers in Ecology and the Environment*, vol 6, pp547–553

Seppälä, R. (2007) 'Global forest sector: Trends, threats and opportunities', in P. H. Freer-Smith, M. S. J. Broadmeadow and J. M. Lynch (eds) *Forestry and Climate Change*, CAB International, Wallingford, pp25–30

Shiyomi, M., and Koizumi, H. (eds.) (2001) *Structure and Function in Agroecosystem Design and Management*, CRC Press, New York

Shukla, A., Fischedick, M. and Dienst, C. (2004) 'Rural electrification in developing countries: Dimensions and trends', in G. Shakravarty (ed) *Renewables and Rural Electrification*, Bibliotheks und Informationssystem der Universität, Oldenburg, pp43–46

UN (2002) *Report of the World Summit on Sustainable Development, Johannesburg, South Africa, 26 August–4 September 2002*, United Nations, New York

URS Forestry (2007) *Australia's Forest Industry in the Year 2020*, Department of Agriculture, Fisheries and Forestry, Canberra, Australia, www.daff.gov.au/__data/assets/pdf_file/0009/643743/2020-report-final.pdf, accessed 5 March 2010

Uryu, Y., Mott, C., Foead, N., Yulianto, K., Budiman, A., Setiabudi, Takakai, F., Nursamsu, Sunarto, Purastuti, E., Fadhli, N., Hutajulu, C. M. B., Jaenicke, J., Hatano, R., Siegert, F. and Stüwe, M. (2008) *Deforestation, Forest Degradation, Biodiversity Loss and CO$_2$ Emissions in Riau, Sumatra, Indonesia*, WWF Indonesia

van der Meer, P. J., Jorritsma, I. T. M. and Kramer, K. (2002) 'Assessing climate change effects on long-term forest development: Adjusting growth, phenology, and seed production in a gap model', *Forest Ecology and Management*, vol 162, pp39–52

Van Eupen, M., Sedze Puchol, T., Sharma, S. and Vijayanand, D. (2007) 'Modelling the distribution of goods and services at the landscape level', NETFOP report no 4 (NETworking FOrest Plantations in a crowded world: Optimizing ecosystem services through improved planning and management)

Vertessy, R. A., Hatton, T. J., Benyon, R. G. and Dawes, W. R. (1996) 'Long-term growth and water-balance predictions for a mountian ash (*Eucalyptus regnans*) forest catchment subject to clearfelling and regeneration', *Tree Physiology*, vol 16, pp221–232

Vitousek, P. M. and others (1997) 'Human alteration of the global nitrogen cycle: Sources and consequences', *Ecological Applications*, vol 7, pp737–750

Wang, C. Y., van der Meer, P. J., Peng, M. C., Douven, W., Hessel, R. and Dang, C. L. (2009) 'Ecosystem services assessment of two watersheds of Lancang River in Yunnan, China, with a decision tree approach', *Ambio*, vol 38, pp47–54

Wunder, S. (2005) *Payments for Environmental Services: Some Nuts and Bolts*, CIFOR Occasional Paper No. 42, CIFOR, Bogor, Indonesia

Wunder, S., Engel, S. and Pagiola, S. (2008) 'Taking stock: A comparative analysis of payments for environmental services programs in developed and developing countries', *Ecological Economics*, vol 65, 834–852

Glossary

Term or Concept	Definition/Explanation	Source
Additionality	Additionality occurs when the amount of CO_2 emission reduced or sequestered within project boundaries is additional to that which would occur without the project, or under business as usual conditions.	Murray et al, 2007
Afforestation	Conversion from other land uses into forest, *or* the increase of the canopy cover to above the 10 per cent threshold.	FAO 2000a
Agroforestry (system)	Intentional growing of trees and shrubs in combination with crops or forage.	NRCS 1996
Agroforestry plantation	Managed stands or assemblages of trees established in an agricultural matrix for subsistence or local sale and for their benefits on agricultural production; usually regular and wide spacing or row planting.	CIFOR 2001
Albedo	The fraction of solar radiation reflected by a surface or object, often expressed as a percentage.	IPCC 2010
Allelopathy	Allelopathy refers to the beneficial or harmful effects of one plant on another plant, both crop and weed species, by the release of chemicals from plant parts by leaching, root exudation, volatilization, residue decomposition and other processes in both natural and agricultural systems.	UF-IFAS 2010

Amenity	Although not explicitly defined by FAO, it is something that conduces comfort, convenience, or enjoyment, which is relative to depending upon socio-cultural contexts (e.g. subjective perspective on an open view of a forested mountain range vs. that of a clear-cut).	Merriam-Webster 2010
Artificial regeneration	A forest crop raised artificially, either by sowing or planting, but no natural regeneration.	FAO 2003a
Benefit transfer	A practice used to estimate economic values for ecosystem services by transferring information available from studies already completed in one location or context to another. This can be done as a unit value transfer or a function transfer.	OECD 2010
Biodiversity	The variability among living organisms from all sources including, inter alia, terrestrial, marine and other aquatic ecosystems and the ecological complexes of which they are part; this includes diversity within species, between species and of ecosystems.	CBD 1992
Biodiversity corridors	They are defined as areas of suitable habitat that provide functional linkages between habitat areas.	IUCN 2009
Bioenergy	Energy from biofuels.	FAO 2008a
Biofuel	Fuel produced directly or indirectly from biomass.	FAO 2008a
Biosolids	Solid, semi-solid, or liquid residues generated during primary, secondary, or advanced treatment of domestic sanitary sewage through one or more controlled processes, such as anaerobic digestion, aerobic digestion, and lime stabilization.	Goverment of Michigan (US) 2010
Canopy stratification	The formation of vertical crowns layers in forest stands.	
Cap	An upper limit imposed on a flux or rate; herein referring to carbon emissions.	IPCC 2010

Carbon mitigation	An intervention to decrease the sources or improve the sinks of carbon-based greenhouse gases.	FAO 2010
Carbon sequestration	The process of removing carbon from the atmosphere and depositing it in a reservoir.	UNFCCC 2010
Carbon sink	Any process, activity or mechanism which removes a carbon-based greenhouse gas, an aerosol or a carbon-based precursor of a greenhouse gas from the atmosphere. Forests and other vegetation are considered sinks because they remove carbon dioxide through photosynthesis.	UNFCCC 2010
Carbon source	Any process, activity, or mechanism that releases a greenhouse gas, an aerosol, or a precursor of a greenhouse gas or aerosol into the atmosphere.	FAO 2010
Carbon stock	A component of the climate system, other than the atmosphere, that has the capacity to store, accumulate or release carbon (including its forms of greenhouse gases). Oceans, soils and forests are examples of carbon stocks or reservoirs.	IPCC 2010
Carbon tax	A carbon tax is an instrument of environmental cost internalization. It is an excise tax on the producers of raw fossil fuels based on the relative carbon content of those fuels.	OECD 2010
Catchment (area)	An area that collects and drains rainwater.	IPCC 2010
Clear-cut (or Clear-fell)	A stand in which essentially all trees have been removed in one operation.	Helms 1998
Commodity markets	Markets where raw or primary products (or goods and services normally intended for sale on the market at a price that is designed to cover their cost of production) are exchanged.	OECD 2010
Community-based approach	An approach involving various sections of the community, drawing on multiple local resources to address a community issue.	–

Compaction	Compaction is the process whereby the density of soils is increased.	OECD 2010
Conservation easement	A voluntary, legally binding agreement that limits certain types of uses or prevents development from taking place on a piece of property now and in the future (vis-à-vis a transfer of usage rights), while protecting the property's ecological or open-space values.	TNC 2010
Contour tilling	Tilling while following the natural contours of the land.	NRCS 2006
Credit programme	A programme designed to supply an amount for which there is a specific obligation of repayment (e.g. loans, trade credits, bonds, bills, etc., and other agreements which give rise to specific obligations to repay over a period of time usually, but not always, with interest.	OECD 2010
Deforestation	Forest conversion to another land use *or* the long-term reduction of tree canopy cover below the 10 per cent threshold.	FAO 2000
Degradation	Degradation is the reduction or loss of the biological or economic productivity resulting from natural processes, land uses or other human activities and habitation patterns such as land contamination, soil erosion and the destruction of the vegetation cover.	OECD 2010
Denitrification	The reduction of nitrate (NO_3^-) or nitrite (NO_2^-) to gaseous N oxides and molecular N_2.	FAO 2001a
Disturbance	Any relative discrete event in time that disrupts ecosystems, community, or population structure and changes resources, substrate availability, or the physical environment.	Helms 1998
Early successional species	Tree species that grow best in forests that have just begun to undergo succession, shortly following a major disturbance.	–

Ecolabelling	A voluntary method of certification of environmental quality (of a product) and/or environmental performance of a process based on life-cycle considerations and agreed sets of criteria and standards.	FAO 2003b
Ecological insurance	The idea that if some species are lost others are available to fill their functional roles, due to functional redundancy (closely linked to biodiversity).	EFTEC 2005
Economical maturity	In the context of its use here, economic maturity is the period of time (or the end of that period) until the most economic benefit can be obtained from harvesting.	OECD 2010
Ecosystem approach	A strategy for the integrated management of land, water and living resources that promotes conservation and sustainable use in an equitable way. It is based on the application of appropriate scientific methodologies focused on levels of biological organization which encompass the essential processes, functions and interactions among organisms and their environment, while recognizing that humans, with their cultural diversity, are an integral component of ecosystems.	CBD 2009
Ecosystem functioning	Capacity of natural processes and components to provide goods and services that satisfy human needs, directly or indirectly.	–
Ecosystem goods and services	The benefits humans derive from ecosystem functions.	MEA 2005
Endemic	Restricted to a particular area: used to describe a species or organism that is confined to a particular geographical region, for example, an island or river basin.	WWF 2010a

Environmental governance	A complex system in which social, economic and political organization of interacting and interdependent groups and organs, public and private, connected by doctrines, ideas and principles, intend to serve a common purpose: the regulation of the use of natural resources.	–
Environmental plantations	Managed forest stand, established primarily to provide environmental stabilization or amenity value, by planting or/and seeding in the process of afforestation or reforestation, usually with even age class and regular spacing.	CIFOR 2001
Evapotrans-piration	The combined process of water evaporation from the Earth's surface and transpiration from vegetation.	IPCC 2010
Farm bunds	An embankment of soil for protection or nutrient-purposes (e.g. planting N-fixing tree species on bunds) of crops.	–
Fast-growing timber plantations	Plantations with growth rates of >15 m³/ha/year and rotations of <20 years.	Cossalter and Pye-Smith 2003
Flagship species	Iconic species that provide a focus for raising awareness and stimulating action and funding for broader conservation efforts (e.g. polar bear for climate change mitigation).	WWF 2010b
Forest certification	A number of market-based instruments that link trade in forest products to the sustainable management of the forest resource, by providing buyers with information on the management standards of the forests from which the timber came.	FAO 2001b
Fossil fuel substitution	The partial, gradual or complete replacement of fossil fuels as a source for meeting energy demands.	–
Fuel load	The amount of fuel available to burn (depends on the type of vegetation, how much of it there is, its 'fineness' and its moisture content).	AAGD 2009

Fuelwood	Wood in the rough (from trunks, and branches of trees) to be used as fuel for purposes such as cooking, heating or power production.	FAO 2001c
Greenhouse gas emissions	Gaseous constituents of the atmosphere, both natural and anthropogenic, that absorb and emit radiation at specific wavelengths within the spectrum of infrared radiation emitted by the Earth's surface, the atmosphere and clouds. This property causes the greenhouse effect. Water vapour (H_2O), carbon dioxide (CO_2), nitrous oxide (N_2O), methane (CH_4) and ozone (O_3) are the primary greenhouse gases in the Earth's atmosphere, and there are a number of entirely human-made greenhouse gases in the atmosphere, such as the halocarbons and other chlorine- and bromine-containing substances.	FAO 2010a
Groundwater	Water stored underground in rock crevices and in the pores of geologic materials that make up the Earth's crust.	USGS 2009
Groundwater recharge	The process by which external water is added to the zone of saturation of an aquifer, either directly into a formation or indirectly by way of another formation.	IPCC 2010
Habitat suitability	The suitability (capacity) of an area to supply the same requisites as a place or type of site where an organism or population naturally occurs.	WWF 2010a
Heterarchy	Networks of elements in which each element shares the same 'horizontal' position of power and authority, each playing a theoretically equal role.	–
Homegardens	Intensively managed areas around houses consisting of trees cultivated together with annual and perennial agricultural crops and livestock.	–

Indigenous community	Residence within or attachment to geographically distinct traditional habitats, ancestral territories, and their natural resources; maintenance of cultural and social identities, and social, economic, cultural and political institutions separate from mainstream or dominant societies and cultures; descent from population groups present in a given area, most frequently before modern states or territories were created and current borders defined; and self-identification as being part of a distinct indigenous cultural group, and the desire to preserve that cultural identity.	IPCC 2010
Industrial plantation	Intensively managed forest stands managed to provide material for sale locally or outside the immediate region.	CIFOR 2001
Industrial roundwood	All industrial wood in the rough (sawlogs and veneer logs, pulpwood and other industrial roundwood) and, in the case of trade, chips and particles and wood residues.	FAO 1997
Infiltration	Flow of water from the land surface into the subsurface.	USGS 2009
Intensive forest management	A regime of forest management, where silvicultural practices define the structure and composition of forest stands. A formal or informal forest management plan exists. A forest is not under intensive management, if mainly natural ecological processes define the structure and composition of stands.	FAO 2004
Interception	In hydrologic terms, the process by which precipitation is caught and held by foliage, twigs and branches of trees, shrubs and other vegetation, and lost by evaporation, never reaching the surface of the ground. Interception equals the precipitation on the vegetation minus streamflow and throughfall.	NOAA 2009
Invasive	Any species that is non-native to a particular ecosystem and whose introduction and spread causes, or is likely to cause, socio-cultural, economic or environmental harm or harm to human health.	FAO 2009

Late successional species	Tree species that grow best in forests that have undergone succession and developed without major disturbances for a long time.	Puettman et al, 2009
Leakage	A negative or counteracting effect on carbon emission reduction efforts due to an increase of emissions in a non-constrained part of a system (i.e. a plan, an area etc.).	IPCC 2010
Lifetime (of a product)	The time length in which a product keeps its utilitarian value or avoids decomposition (and carbon emission) associated with discard.	–
Livelihood services	Human services that increase the standard of living or give the tools to do so (e.g. exchange of workload, division of tasks, support in the case of emergencies as well as employment opportunities and temporary provision of land to landless poor).	
Lopping	The cutting off of branches, usually for fodder or fuelwood.	–
Managed secondary forests (with planting)	Managed forest, where forest composition and productivity is maintained through additional planting or/and seeding.	–
Market-based approach	An approach using price mechanisms (e.g. taxes and auctioned tradable permits), among other instruments, to achieve a goal.	IPCC 2010
Marrakesh Accords	Agreements reached at COP-7 which set various rules for 'operating' the more complex provisions of the Kyoto Protocol. Among other things, the accords include details for establishing a greenhouse-gas emissions trading system; implementing and monitoring the Protocol's Clean Development Mechanism; and setting up and operating three funds to support efforts to adapt to climate change.	UNFCCC 2010
Mass wasting	Soil movement such as landslips, slumps and debris flows (landslides).	FAO 2008b

Monoculture	A forest stand composed of a single tree species.	Puettman et al, 2009
Native forest	Forests composed of indigenous tree species	–
Natural disturbance	See Disturbance.	–
Natural regeneration	Re-establishment of a forest stand by natural means, i.e. by natural seeding or vegetative regeneration. It may be assisted by human intervention, e.g. by scarification of the soil or fencing to protect against wildlife or domestic animal grazing.	FAO 2007
Naturalness	It is the degree to which ecosystems resemble natural, not anthropogenically modified, ecosystems in terms of their structure, function and composition.	–
Niche complementarity	The ability of different species that use different resources, or the same resources but at different times or different points in space, to fill the available limited space to a greater extent than a single species.	–
Non-point pollution	The source(s) of pollution whose discrete origins are difficult to accurately pin down.	FAO 2003c
Non-timber forest products	Products of biological origin, other than wood, derived from forests, other wooded land and trees outside forests. NTFP may be gathered from the wild, or produced in forest plantations, agroforestry schemes and from trees outside forests.	FAO 2008c
Non-wood forest products	see Non-timber forest products.	–
Old-growth forest	Ecosystems distinguished by old trees and related structural attributes. Old growth encompasses the later stages of stand development that typically differ from earlier stages in a variety of characteristics which may include tree size, accumulations of large dead woody material, number of canopy layers, species composition, and ecosystem function.	FAO 2002

Outgrower scheme	A contractual partnership between growers or landholders and a company for the production of commercial forest products.	FAO 2001d
Overyielding effect	When the production of biomass in species mixtures exceeds the productivity expected on the basis of the yields of the contributing species when grown in a monoculture.	–
Peak flow	The maximum instantaneous discharge of a stream or river at a given location.	USGS 2009
Permanence	The time length of which a particular plan, scheme or action can ensure carbon sequestration without reversing effects (e.g. planted trees that will later be harvested for wood-burning would offset the initial carbon sequestration and the permanence in this case would be analogous with rotation period until harvest).	–
Plantations	Forests of introduced species and in some cases native species, established through planting or seeding, with few species, even spacing and/or even-aged stands.	FAO 2004
Planted forests (contrast with 'Plantations')	Forests predominantly composed of trees established through planting and/or after deliberate seeding of native or introduced species.	Carle and Holmgren, 2008
Polyculture	Stand with more than one tree species.	–
Polyphagous	Feeding on different types of food (e.g. a pest population that feeds on different host species).	–
Primary forest	A forest of native species, where ecological processes are undisturbed by human activities.	FAO 2004
Production forest	A forest actually designated for production of forest goods, i.e. where the extraction of forest products, usually wood and fibre, is the predominant management objective, including both wood and non-wood forest products.	FAO 2006

Productive plantation	Forest plantations predominantly intended for the provision of wood, fibre and non-wood products.	FAO 2006
Protective plantation	Forest plantations predominantly for the provision of services such as protection of soil and water, rehabilitation of degraded lands, combating desertification, etc.	FAO 2006
Provisioning ecosystem services	Products obtained from ecosystems (e.g. food, fresh water, fuelwood, fibre, biochemicals, genetics etc.).	FAO 2010b
Reforestation	Reforestation is the re-establishment of forest formations after a temporary condition with less than 10 per cent canopy cover due to human-induced or natural perturbations.	FAO 2000
Regulation ecosystem services	Benefits obtained from the regulation of ecosystems (e.g. climate regulations, disease regulation, water regulation and purification, pollination etc.).	FAO 2010b
Residence time	With respect to greenhouse gases, residence time usually refers to how long a particular molecule remains in the atmosphere.	EPA 2009
Resilience	The capacity of an ecosystem to return to the pre-condition state following a perturbation, including maintaining its essential characteristics, taxonomic composition, structures, ecosystem functions, and process rates.	Holling 1973
Resistance	The capacity of the ecosystem to absorb disturbances and remain largely unchanged.	Holling 1973
Restorability	The possibility of spontaneous renewal or human-aided restoration of ecosystems.	–
Rotation length	Period between regeneration establishment and final cutting of even-aged forests.	Helms, 1998
Run-off	That part of the precipitation, snow melt, or irrigation water that appears in uncontrolled surface streams, rivers, drains or sewers.	USGS 2009

Scalping	Removing vegetation and other organic or inorganic material to expose underlying mineral soil and prepare an area for planting or seeding.	Helms 1998
Silviculture	The art and science of producing and tending a forest to achieve the objectives of management.	Puettman et al, 2009
Site preparation	Any treatment of a forest site to prepare the soil for the establishment of a new crop of trees by either plantation or natural means.	Puettman et al, 2009
Smallholders	People living in rural areas who owns small areas of land that he/she cultivates for subsistence or commercial purposes relying principally on family labour.	–
Social forestry	Programmes are specifically geared to enable a social actor to expand production to meet supply needs.	FAO 1992
Species diversity	The total number and distribution of species in an area: includes components of both species richness and the amount of evenness or dominance among the species present (contrast with Species richness).	FAO 2010c
Species richness	The total number of species in an area measured by a standard protocol (not to be confused with Species Diversity)	FAO 2010c
Stakeholder	Person or entity holding grants, concessions, or any other type of value that would be affected by a particular action or policy.	FAO 2010a
Supporting ecosystem services	Services necessary for the production of all other ecosystem services (e.g. soil formation, nutrient cycling and primary production).	FAO 2010b
Sustainability	Characteristic by which a process or state can be maintained at a certain level indefinitely. In the environmental context, the term refers to the potential longevity of vital human ecological support systems and the various systems on which they depend in balance with the impacts of our unsustainable or sustainable design.	Meyers (in press)

Sustainable forest management	The stewardship and use of forests and forest lands in a way, and at a rate, that maintains their biological diversity, productivity, regeneration capacity, vitality and their potential to fulfil, now and in the future, relevant ecological economic and social functions, at local, national and global levels, and that does not cause damage on other ecosystems.	FAO 2000b
Thinning	Partial removal of trees in an immature stand to select for a specific species or to increase the growth rate of the remaining trees.	Puettman et al, 2009
Traditional community	Community that has a long tradition of land use with typically common access to forest areas in remote, still forested landscapes and with a low level of infrastructure.	–
Transfer of technology (Technology transfer)	The broad set of processes covering the exchange of knowledge, money and goods amongst different stakeholders that lead to the spreading of technology for adapting to or mitigating climate change.	IPCC 2010
Transpiration	The evaporation of water vapour from the surfaces of leaves through stomata.	IPCC 2010
Uniqueness	A measure of relative difference for a species from others.	–
Vapour flux	The flow rate of a gaseous substance.	–
Water quality	A term used to describe the chemical, physical, and biological characteristics of water, usually in respect to its suitability for a particular purpose.	USGS 2009
Water regulation	The control of water flow (velocity and volume).	–
Weir	A small overflow-type dam commonly used to raise the level of a river or stream.	–

References

AAGD (Australian Attorney-General's Department) (2009) 'Australian emergency management terms thesaurus', http://library.ema.gov.au/emathesaurus/, accessed 24 March 2010

Carle, J. and Holmgren, P. (2008) 'Wood from planted forests: A global outlook 2005–2030', *Forest Products Journal*, vol 58, no 12

CBD (1992) *Convention on Biological Diversity*, www.cbd.int/convention/convention.shtml, accessed 16 March 2010

CBD (2009) 'Ecosystem approach', *Convention on Biological Diversity*, www.cbd.int/ecosystem/, accessed 24 March 2010

CIFOR (2001) 'Typology of Planted Forests', *CIFOR InfoBrief*, Center for International Forestry Research (CIFOR)

Cossalter, C. and Pye-Smith, C. (2003) 'Fast-wood forestry: myths and realities', *Forest Perspectives*, CIFOR, Indonesia

EFTEC (2005) 'The economic, social and ecological value of ecosystem services: a literature review', Economics for the Environment Consultancy, www.cbd.int/doc/case-studies/inc/cs-inc-uk6-en.pdf, accessed 24 March 2010

EPA (2009) 'Glossary of climate change terms', United States Environmental Protection Agency, www.epa.gov/climatechange/glossary.html#P, accessed 24 March 2010

FAO (1992) 'Community forestry: Ten years in review', CF Note 7, www.fao.org/docrep/u5610e/u5610e00.htm#Contents, accessed 16 March 2010

FAO (1997) 'Asia-Pacific forestry sector outlook study: Trends and outlook for forest products consumption, production and trade in the Asia-Pacific region', Working Paper No. APFSOS/WP/12, Food and Agriculture Organization, www.fao.org/DOCREP/W7705E/w7705e00.htm#Contents, accessed 16 March 2010

FAO (2000a) 'Definitions of forest change processes', Working Paper No. 33, Food and Agriculture Organization, Forest Resources Assessment, www.fao.org/docrep/006/ad665e/ad665e04.htm#TopOfPage, accessed 16 March 2010

FAO (2000b) 'Asia-Pacific Forestry Commission: Development of national-level criteria and indicators for the sustainable management of dry forests of Asia', Workshop Report, RAP Publication 2000/07, Food and Agriculture Organization, www.fao.org/docrep/003/x6896e/x6896e00.htm#Contents, accessed 12 March 2010

FAO (2001a) 'Global estimates of gaseous emissions of NH_3, NO and N_2O from agricultural land', Y2780/E, Food and Agriculture Organization, www.fao.org/docrep/004/y2780e/y2780e00.htm#P-1_0, accessed 16 March 2010

FAO (2001b) 'Criteria and indicators of sustainable forest management of all types of forests and implications for certification and trade', COFO-2001/3, Food and Agriculture Organization, Committee on Forestry, 15th Session, www.fao.org/docrep/MEETING/003/X8783E.HTM#P82_8256, accessed 15 March 2010

FAO (2001c) 'Unified wood energy terminology', J0926/E, Food and Agriculture Organization, www.fao.org/docrep/008/j0926e/J0926e03.htm#TopOfPage, accessed 16 March 2010

FAO (2001d) 'Forestry out-grower schemes: A global view', Forest Plantations Thematic Papers, Working Paper FP/11, www.fao.org/docrep/004/ac131e/ac131e03.htm#bm3.1, accessed 12 March 2010

FAO (2002) 'Second expert meeting on harmonizing forest-related definitions for use by various stakeholders', Y4171/E, Food and Agriculture Organization, www.fao.org/docrep/005/y4171e/y4171e00.htm#TopOfPage, accessed 12 March 2010

FAO (2003a) 'Planted forests database (PFDB): Structure and contents', Working Paper FP/25, Food and Agriculture Organization

FAO (2003b) 'The ecosystem approach to fisheries', Y4773/E, Food and Agriculture Organization, www.fao.org/DOCREP/006/Y4773E/ y4773e00.htm#Contents, accessed 16 March 2010

FAO (2003c) 'Preparing national regulations for water resources management principles and practice', FAO Legislative Study, Y5051/E, 80, Food and Agriculture Organization, www.fao.org/DOCREP/006/Y5051E/ y5051e00.htm#Contents, accessed 16 March 2010

FAO (2004) 'Global forest resources assessment update 2005 Terms and definitions', Forest Resources Assessment WP 83, Food and Agriculture Organization, www.fao.org/docrep/007/ae156e/AE156E00.htm#TopOfPage, accessed 12 March 2010

FAO (2006) 'Responsible management of planted forests: Voluntary guidelines', Planted Forests and Trees Working Paper FP37E, Food and Agriculture Organization, www.fao.org/docrep/009/j9256e/J9256E00.htm#TopOfPage, accessed 12 March 2010

FAO (2007) 'Definitional issues related to reducing emissions from deforestation in developing countries', Forests and Climate Change: Working Paper 5, Food and Agriculture Organization, ftp://ftp.fao.org/docrep/FAO/009/j9345e/j9345e00.pdf, downloaded 15 March 2010

FAO (2008a) 'Bioenergy: Terminology', Food and Agriculture Organization, www.fao.org/bioenergy/52184/en/, accessed 16 March 2010

FAO (2008b) 'Forests and water: A thematic study prepared in the framework of the Global Forest Resources Assessment 2005', FAO Forestry Paper 155, Food and Agriculture Organization, www.fao.org/docrep/011/i0410e/i0410e00.htm, downloaded 15 March 2010

FAO (2008c) 'What are non-wood forest products?' Food and Agriculture Organization www.fao.org/forestry/nwfp/6388/en/, accessed 12 March 2010

FAO (2009) 'Invasive species: impacts on forests and forestry', Food and Agriculture Organization, www.fao.org/forestry/aliens/en/, accessed 12 March 2010

FAO (2010a) 'Climate change terminology', Food and Agriculture Organization, www.fao.org/climatechange/49365/en/#c, accessed 16 March 2010

FAO (2010b) 'What are ecosystem services?', Food and Agriculture Organization, www.fao.org/ES/ESA/pesal/aboutPES1.html, accessed 15 March 2010

FAO (2010c) 'Species richness', Food and Agriculture Organization, www.fao.org/gtos/tems/variable_show.jsp?VARIABLE_ID=134, accessed 16 March 2010

Government of Michigan (US) (2010) 'Biosolids definition', State of Michigan (United States): Department of Natural Resources and Environment, www.michigan.gov/, accessed 12 March 2010

Helms, J. (1998) *The Dictionary of Forestry*, The Society of American Foresters. Bethesda

Holling, C. S. (1973) 'Resilience and stability of ecosystems', *Annual Review of Ecology and Systematics*, vol 4, pp1–23.

IPCC (2010) 'Glossary', Intergovernmental Panel on Climate Change, www.ipcc.ch/publications_and_data/publications_and_data_glossary.htm, accessed 16 March 2010

IUCN (2009) 'Connectivity conservation: International experience in planning, establishment and managementof biodiversity corridors', International Union for Conservation of Nature, www.iucn.org/, downloaded 12 March 2010

MEA (Millennium Ecosystem Assessment) (2005) *Ecosystems and Human Well-being: Synthesis*, World Resources Institute, Island Press, Washington, DC

Merriam-Webster (2010) 'Dictionary and Thesaurus', www.merriam-webster.com/, accessed 16 March 2010

Meyers, R. (ed) (in press) *Encyclopedia of Sustainability Science and Technology*, Springer

Murray, B. C., Sohngen, B. and Ross, M. T. (2007) 'Economic consequences of consideration of permanence, leakage and additionality for soil carbon sequestration projects', *Climatic Change*, vol 80, pp127–143

NOAA (2009) 'National Weather Service glossary', National Oceanic Atmospheric Administration (United States), www.nws.noaa.gov/glossary/, accessed 24 March 2010

NRCS (1996) 'Agroforestry for farms and ranches', Natural Resources Conservation Service (USDA), www.nrcs.usda.gov/, accessed 16 March 2010

NRCS (2006) 'Contour farming', Conservation Practice Job Sheet PA 330, Natural Resources Conservation Service (USDA), www.nrcs.usda.gov/, accessed 24 March 2010

OECD (2010) 'Glossary of statistical terms', http://stats.oecd.org/, accessed 16 March 2010

Puettman, K. J., Coates, K. D. and Messier, C. (2009) *A Critique of Silviculture: Managing for Complexity*, Island Press, Washington, DC

TNC (2010) 'How we work – Conservation methods: Conservation easements', The Nature Conservancy, www.nature.org/aboutus/howwework/conservationmethods/privatelands/conservationeasements/, accessed 24 March 2010

UF-IFAS (2010) 'Allelopathy: How plants suppress other plants', University of Florida-Institute of Food and Agricultural Sciences, http://solutionsforyourlife.ufl.edu/, accessed 16 March 2010

UNFCCC (2010) 'Glossary of climate change acronyms', United Nations Framework Convention on Climate Change, http://unfccc.int/, accessed 12 March 2010

USGS (2009) 'Water science glossary of terms', United States Geological Survey, http://ga.water.usgs.gov/edu/dictionary.html#G, accessed 24 March 2010

WWF (2010a) 'Biodiversity glossary', World Wide Fund for Nature, www.panda.org/about_our_earth/biodiversity/biodiversity_glossary/, accessed 24 March 2010

WWF (2010) 'Priority species', World Wide Fund for Nature, www.panda.org/what_we_do/endangered_species/, accessed 24 March 2010

Index